AN INTRODUCTION TO THE MANAGEMENT AND REGULATION OF HAZARDOUS WASTE

Emmett B. Moore, Ph.D.
Adjunct Professor of Environmental Science
Washington State University

Library of Congress Cataloging-in-Publication Data

Moore, Emmett Burris, 1929– .
An introduction to the management and regulation of hazardous waste / written by Emmett Moore.
p. cm
Includes index.
ISBN 1-57477-088-8 (soft cover : alk. paper)
1. Hazardous wastes—Management. 2. Hazardous wastes—Law and legislation—United States. I. Title.

TD1030.M66 2000
363.72′8756—dc21

99-057216

Printed in the United States of America

Battelle Press
505 King Avenue
Columbus, Ohio 43201-2693, USA
614-424-6393 or 1-800-451-3543
Fax: 614-424-3819
E-mail: press@battelle.org
Website: www.battelle.org/bookstore

Contents

Preface

This book is a result of my experience as a physical chemist, as a state regulator, as a staff scientist at the Pacific Northwest National Laboratory, and as an adjunct professor of environmental science at Washington State University. All of the disciplines in my past were important to me in preparing this book: chemistry, regulatory affairs, environmental science, and teaching. Indeed, hazardous waste management is very much an interdisciplinary field that requires an understanding of basic chemistry, environmental law, and public participation, as well as many other fields such as biology, geology, and toxicology in order for the practitioner to be successful. This book is an introduction to hazardous waste management. It covers the two major federal laws that regulate hazardous wastes: the Resource Conservation and Recovery Act (RCRA), which regulates active hazardous waste sites and newly generated hazardous wastes, and the Comprehensive Environmental Response, Compensation, and Liability Act (CERCLA), which regulates abandoned or retired hazardous waste sites, and some other things as well. The book also identifies frequently encountered hazardous wastes, particularly those found at CERCLA sites, discusses their chemistry and toxicology, and describes technologies for their removal from the environment. It is my hope that this book will be useful to upper division university students, graduate students, state and federal regulators, and industrial personnel charged with hazardous waste management and/or environmental cleanup.

This book is dedicated with love to my parents, Emmett and Iris Moore; fondly to all of my students, my teachers, and my faculty colleagues at the University of Minnesota and Washington State University; and to all of my regulatory colleagues in the State of Minnesota.

EMMETT MOORE

Acronyms

ARAR	applicable or relevant and appropriate requirement
BDAT	best demonstrated available technology
BOD	biological oxygen demand
BTEX	benzene, toluene, ethyl benzene, xylenes
CAA	Clean Air Act
CAAA	Clean Air Act Amendments of 1990
CEQ	Council on Environmental Quality
CERCLA	Comprehensive Environmental Response, Compensation, and Liability Act
CFC	chlorofluorocarbon
CFR	*Code of Federal Regulations*
CMI	corrective measure implementation
CMS	corrective measure study
COE	U.S. Army Corps of Engineers
CWA	Clean Water Act
DOD	U.S. Department of Defense
DOE	U.S. Department of Energy
DOT	U.S. Department of Transportation
DOI	U.S. Department of the Interior
EA	environmental assessment
EIS	environmental impact statement
EPA	U.S. Environmental Protection Agency
EPCRA	Emergency Planning and Community Right-to-Know Act
ESA	Endangered Species Act
FFCA	Federal Facility Compliance Act

FIFRA	Federal Insecticide, Fungicide, and Rodenticide Act
FR	*Federal Register*
FWS	U.S. Fish and Wildlife Service
HEPA	high energy particulate air
HSWA	Hazardous and Solid Waste Amendments (to RCRA)
IRIS	Integrated Risk Information System
LDR	land disposal restrictions
MCL	maximum contaminant level
MCLG	maximum contaminant level goal
MTBE	methyl tertiary-butyl ether
NAAQS	National Ambient Air Quality Standards
NCP	National Contingency Plan
NEPA	National Environmental Policy Act
NESHAP	National Emission Standards for Hazardous Air Pollutants
NHPA	National Historic Preservation Act
NIMBY	not in my back yard
NMFS	National Marine Fisheries Service
NPDES	National Pollutant Discharge Elimination System
NPL	National Priorities List
NRC	U.S. Nuclear Regulatory Commission
NRDA	natural resource damage assessment
OPA	Oil Pollution Act
OSHA	Occupational Safety and Health Act
PAH	polycyclic aromatic hydrocarbon
PAN	peroxyacetyl-nitrate
PA/SI	preliminary assessment/site inspection
PCB	polychlorinated biphenyl
PCE	perchloroethylene, tetrachloroethylene
PSD	prevention of significant deterioration

RCRA	Resource Conservation and Recovery Act
RD/RA	remedial design/remedial action
RFA	RCRA facility assessment
RI/FS	remedial investigation/feasibility study
RFI	RCRA facility investigation
ROD	record of decision
SARA	Superfund Amendments and Reauthorization Act
SDWA	Safe Drinking Water Act
SHPO	state historic preservation officer
SI	International System of Units
SS	suspended solids
SWDA	Solid Waste Disposal Act
TCE	trichloroethylene
TCLP	toxicity characteristic leaching procedure
TSCA	Toxic Substances Control Act
TSD	treatment, storage, and/or disposal
UIC	underground injection control
USC	*United States Code*
UST	underground storage tank
VOC	volatile organic compound

AN INTRODUCTION TO THE MANAGEMENT AND REGULATION OF HAZARDOUS WASTE

CHAPTER 1

Introduction

The management of air and water pollution has been a major concern in the United States since the end of World War II. Precedent for this concern dates back hundreds of years to when problems existed in Europe with respect to water-borne communicable diseases such as cholera and typhoid and with respect to the health effects of air pollution from burning wood and coal. Major regulatory response to these concerns in the United States began in the 1960s with the passage of the Clean Air Act in 1963, partly in response to air-quality conditions in the Los Angeles basin. At the same time, however, land disposal of wastes received much less attention, probably because of an "out of sight, out of mind" attitude. Nevertheless, many buried wastes contained hazardous solids and liquids that can be toxic to animals and humans when released to the environment. These substances can be poisons, carcinogens, mutagens, and/or teratogens. The release pathway can be through the air if the substance is volatile or through ground water if the substance is soluble in water. Environmental disasters in the United States, such as Love Canal (Ember 1989), Times Beach (Hileman 1997), and the Stringfellow acid pits (Ember 1985, ER 1993), made the "out of sight out of mind" attitude impossible to sustain, and major attention was directed toward managing hazardous wastes. This attention included Congressional action directed at both managing newly created hazardous wastes and cleaning up retired or abandoned waste sites. The two major environmental laws in the United States relating to the management of hazardous wastes and the cleanup of retired or abandoned hazardous waste disposal sites are the Resource Conservation and Recovery Act (RCRA) and the Comprehensive Environmental Response, Compensation, and Liability Act (CERCLA). In addition, other major environmental laws deal with air and water pollution, specifically the Clean Air Act (CAA), the Clean Water Act (CWA), the Oil Pollution Act (OPA), and the Safe Drinking Water Act (SDWA).

In this text, the reader is introduced to the chemical and technical aspects of hazardous waste management, the chemical and technical aspects of air and water pollution control (as related to hazardous waste management), the chemical and technical aspects of the cleanup of soils and ground water contaminated with hazardous wastes, and the regulatory aspects of RCRA and CERCLA. The regulatory aspects of the CAA, CWA, OPA, and SDWA are discussed as they have direct application to RCRA and CERCLA, and the reader is introduced to the chemistry and toxicology of important hazardous substances, the assessment of risk, and the siting of hazardous waste facilities.

RCRA applies to the management of active hazardous waste facilities and CERCLA applies to the cleanup of inactive hazardous waste disposal sites, to the cleanup of hazardous substances released into the environment, and to the assessment of monetary damages for injuries to natural resources caused by releases of hazardous substances to the environment. These acts are discussed fully in the text, but are summarized in the next two sections to set the stage for what follows.

Note: Citations of federal law, federal regulations, and official notification documents of the federal government are given throughout the text as a convenience to the reader. Federal law is codified in the United States Code, abbreviated U.S.C. or sometimes USC. Thus, Section 2 of the National Environmental Policy Act (NEPA) became Section 4321 of Title 42 of the United States Code and may be cited either as NEPA Section 2 or 42 USC 4321. Federal regulations are codified in the *Code of Federal Regulations*, abbreviated CFR. Part 1500 of Title 40 of the *Code of Federal Regulations* is cited as 40 CFR 1500. Official announcements of the federal government related to regulatory matters are printed in the *Federal Register* (FR). Citations of the

Federal Register are by volume number and page number, 51 FR 15625, for example. Sometimes the date of publication is given, as in 51 FR 15625, April 25, 1986.

1.1 **THE RESOURCE CONSERVATION AND RECOVERY ACT** (42 USC 6901 et seq.)

RCRA was passed in 1976 as an amendment to the Solid Waste Disposal Act (SWDA) (enacted in 1965) and was amended in 1984 by the Hazardous and Solid Waste Amendments (HSWA). RCRA is described as a cradle-to-grave system for the management of hazardous waste. It applies mainly to active facilities. RCRA is administered at the federal level by the U.S. Environmental Protection Agency (EPA). Authority to administer RCRA can be delegated to the states by EPA. RCRA requires a manifest system for the generation, transportation, treatment, storage, and disposal of hazardous waste and requires permits for the treatment, storage, or disposal (TSD) of hazardous wastes. A waiver of sovereign immunity appears in RCRA. (A waiver of sovereign immunity, in this case, means that federal facilities may be regulated by appropriate state agencies.) EPA's hazardous waste regulations appear in the *Code of Federal Regulations* in Title 40 Parts 260-281 (40 CFR 260-281). See Chapter 6 for a complete discussion of RCRA.

1.2 **THE COMPREHENSIVE ENVIRONMENTAL RESPONSE, COMPENSATION, AND LIABILITY ACT** (42 USC 9601 et seq.)

CERCLA was passed in 1980 and amended by the Superfund Amendments and Reauthorization Act (SARA) in 1986. CERCLA and SARA provide for remedial action at inactive or abandoned hazardous waste disposal sites (cleanup), provide for removal (also cleanup) of spills of hazardous substances, provide for reporting releases to the environment of hazardous substances, and provide for natural resource damage assessments. Significantly, no permits are required for CERCLA cleanup activities conducted entirely on the CERCLA site. A waiver of sovereign immunity appears in CERCLA. EPA's CERCLA regulations appear in 40 CFR 300-302. See Chapter 7 for a complete discussion of CERCLA.

1.3 SOLID WASTES, HAZARDOUS WASTES, AND HAZARDOUS SUBSTANCES

It is important at the outset to understand the legal definitions of "solid wastes," "hazardous wastes," and "hazardous substances." An understanding of these terms is necessary to an understanding of RCRA and CERCLA. Solid wastes and hazardous wastes are defined under RCRA and hazardous substances are defined under CERCLA. A hazardous waste must first be a solid waste under RCRA, and hazardous substances under CERCLA include RCRA hazardous wastes. The legal definitions in RCRA and CERCLA are somewhat generic. They are, however, made specific in the regulations.

1.3.1 Definition of Solid Wastes under RCRA

"Solid waste" is defined under RCRA Sec. 1004(27) as follows:

> The term "solid waste" means any garbage, refuse, sludge, from a waste treatment plant, water supply treatment plant, or air pollution control facility and other discarded material, including solid, liquid, semisolid, or contained gaseous material resulting from industrial, commercial, mining, and agricultural operations, and from community activities, but does not include solid or dissolved material in domestic sewage, or solid or dissolved materials in irrigation return flows or industrial discharges which are point sources subject to permits under section 402 of the Federal Water Pollution Control Act, as amended (86 Stat. 880), or source, special nuclear, or byproduct material as defined by the Atomic Energy Act of 1954, as amended (68 Stat. 923).

Note that a *solid* waste under RCRA includes liquids and contained gaseous material. There are few chemists in Congress.

1.3.2 Definition of Hazardous Wastes under RCRA

"Hazardous waste" is defined under RCRA Sec. 1004(5) as follows:

> The term "hazardous waste" means a solid waste, or combination of solid wastes, which because of its quantity, concentration, or physical, chemical or infectious charac-

> teristics may (A) cause, or significantly contribute to an increase in mortality or an increase in serious irreversible, or incapacitating reversible, illness; or (B) pose a substantial present or potential hazard to human health or the environment when improperly treated, stored, transported, or disposed of, or otherwise managed.

This definition is, of course, generic and not too helpful if one has a barrel of material that needs to be identified and characterized for its potential as a hazardous waste. The EPA regulations in 40 CFR 261 are more helpful.

A hazardous waste must first be a solid waste. Specifically, a hazardous waste is a solid waste which, in brief, is injurious to human health or a hazard to the environment. Hazardous wastes include listed wastes and characteristic wastes. Listed wastes are listed in 40 CFR 261 either by chemical name or as a named process stream. Characteristic wastes are those that are "reactive, corrosive, toxic, or ignitable." Characteristic wastes are not necessarily listed by name in 40 CFR 261. Most radioactive substances, i.e., source, special nuclear, and byproduct materials, are excluded from regulation under RCRA. Again, in order to be a hazardous waste under RCRA, a waste must first be a solid waste, and a solid waste under RCRA, as noted above, may be a liquid, solid, or contained gas.

1.3.3 **Definition of Hazardous Substances under CERCLA**

"Hazardous substance" is defined under CERCLA Sec. 101(14) as follows:

> The term "hazardous substance" means (A) any substance designated pursuant to section 311(b)(2)(A) of the Federal Water Pollution Control Act [CWA], (B) any element, compound, mixture, solution, or substance designated pursuant to section 102 of this Act, (C) any hazardous waste having the characteristics identified under or listed pursuant to section 3001 of the Solid Waste Disposal Act [RCRA] (but not including any waste the regulation of which under the Solid Waste Disposal Act has been suspended by Act of Congress), (D) any toxic pollutant listed under section 307(a) of the Federal Water Pollution

Control Act [CWA], (E) any hazardous air pollutant listed under section 112 of the Clean Air Act, and (F) any imminently hazardous chemical substance or mixture with respect to which the Administrator has taken action pursuant to Section 7 of the Toxic Substances Control Act [TSCA]. The term does not include petroleum, including crude oil or any fraction thereof which is not otherwise specifically listed or designated as a hazardous substance under subparagraphs (A) through (F) of this paragraph and the term does not include natural gas, natural gas liquids, liquefied natural gas, or synthetic gas usable for fuel (or mixtures of natural gas and such synthetic gas).

Hazardous substances as defined under CERCLA are listed in 40 CFR 302. CERCLA hazardous substances include RCRA hazardous wastes, as well as substances defined as hazardous under the CAA, CWA, and Toxic Substances Control Act (TSCA). Because radionuclides are listed as hazardous air pollutants under the CAA, they are regulated under CERCLA and are therefore subject to CERCLA cleanup requirements. But note that oil and natural gas liquids are not hazardous substances under CERCLA.

1.4 SOME OTHER IMPORTANT DEFINITIONS

Other important legal terms in hazardous waste management include "disposal," "environment," "natural resources," and "pollution" or "pollutant," which are defined below. These definitions are presented here not because they are crucial to the understanding of the management of hazardous waste, but rather because they show the legalistic distinctions that are often made in laws and regulations. These distinctions are sometimes more confusing than helpful to the person interested in understanding the management of hazardous waste.

1.4.1 Disposal

The definition of "disposal" is the same under both RCRA (Sec. 1004(3)) and CERCLA (Sec. 101(29). CERCLA adopts the RCRA definition, which is:

> The term "disposal" means the discharge, deposit, injection, dumping, spilling, leaking, or placing of any solid waste or hazardous waste into or on any land or water so that such solid waste or hazardous waste or any constituent thereof may enter the environment or be emitted into the air or discharged into any water, including ground water.

1.4.2 **The Environment**

For the purposes of environmental science, the environment includes the air; surface waters; ground waters; the land, both surface and subsurface; and the ecosystem, which includes the entire biological universe. Note, however, that the term "environment" may be defined differently in various federal laws and may not include the ecosystem. For example, "environment" is not defined in RCRA, and the definition of the environment under CERCLA does not include the ecosystem (CERCLA Sec. 101(8)):

> The term "environment" means (A) the navigable waters, the waters of the contiguous zone, and the ocean waters of which the natural resources are under the exclusive management authority of the United States under the Magnuson Fishery Conservation and Management Act, and (B) any other surface water, ground water, drinking water supply, land surface or subsurface strata, or ambient air within the United States or under the jurisdiction of the United States.

1.4.3 **Natural Resources**

On the other hand, the definition of "natural resources" under CERCLA does include the ecosystem (CERCLA Sec. 101(16)):

> The term "natural resources" means land, fish, wildlife, biota, air, water, ground water, drinking water supplies, and other such resources belonging to, managed by, held in trust by, appertaining to, or otherwise controlled by the United States (including the resources or fishery conservation zone established by the Magnuson Fishery Conservation and Management Act), any State or local gov-

> ernment, any foreign government, any Indian tribe, or, if such resources are subject to a trust restriction on alienation, any member of an Indian tribe.

1.4.4 **Pollution, Pollutant**

"Pollution" is defined under the Clean Water Act as follows (CWA Sec. 502(19)):

> The term "pollution" means the man-made or man-induced alteration of the chemical, physical, biological, or radiological integrity of the water.

The reader is also advised that specific definitions of the following terms appear in federal law as noted:

- "hazardous air pollutant," CAA Sec. 112(a)(6),
- "pollutant," CWA Sec. 502(6), and
- "toxic pollutant," CWA Sec. 502(13).

1.4.5 **Extremely Hazardous Substances**

In case these terms are not enough, the reader also needs to be aware that "extremely hazardous substances" appear in a stand-alone section of SARA called the Emergency Planning and Community Right-To-Know Act (EPCRA):

- "extremely hazardous substance," EPCRA Sec. 302(a)(2).

1.5 **MOORE'S LAWS AND OPINIONS**

Several principles, which are not necessarily original with the author, should be kept in mind when studying environmental science and hazardous waste management. These principles make it easier to consider hazardous waste management within the context of the world environment.

1.5.1 **Moore's Laws of Environmental Science**

1. *"All things are connected."*
   "All things are connected" is attributed to Chief Seattle in his speech near Seattle, WA, in 1854 marking the transfer of ancestral Indian lands to the federal government. For our purpose, it means that all things on earth are connected and that anything that goes into the air, the water, or the ground can migrate to other media and can reach us, plants, and other animals. The statement is an introduction to pathway analysis (Sec. 3.2). The statement is also true of things legal, as it is of things environmental.

2. *Everything on earth should operate in cycles.*
   The earth maintains its liveability by supporting various cycles such as the water cycle (Figure 1), carbon cycle, nitrogen cycle, sulfur cycle, and various biological cycles. When man intrudes on one of these cycles, the cycle can be disrupted with eventual unpleasant results. An example of this is the carbon cycle in which the air and the oceans are not now able to assimilate all of the carbon dioxide that results from fossil fuel burning. The concentration of carbon dioxide is thus increasing in the atmosphere and a warming (possibly a disastrous warming) of the earth's atmosphere is predicted to result from the greenhouse effect engendered by the presence of excess carbon dioxide in the atmosphere. (See Sec. 3.1.4.) Another example is solid waste buried in areas where ground water can contact the waste form. If the wastes are soluble, then they will move with the ground water to wells, streams, lakes, and eventually the oceans. It has been suggested that wastes (such as carbon dioxide and solid wastes) do not accumulate when humans are not present (Herrick 1994). If so, then it is humans that operate outside of the various established chemical and biological cycles. Continued and flagrant operation outside nature's established cycles is clearly a prescription for disaster.

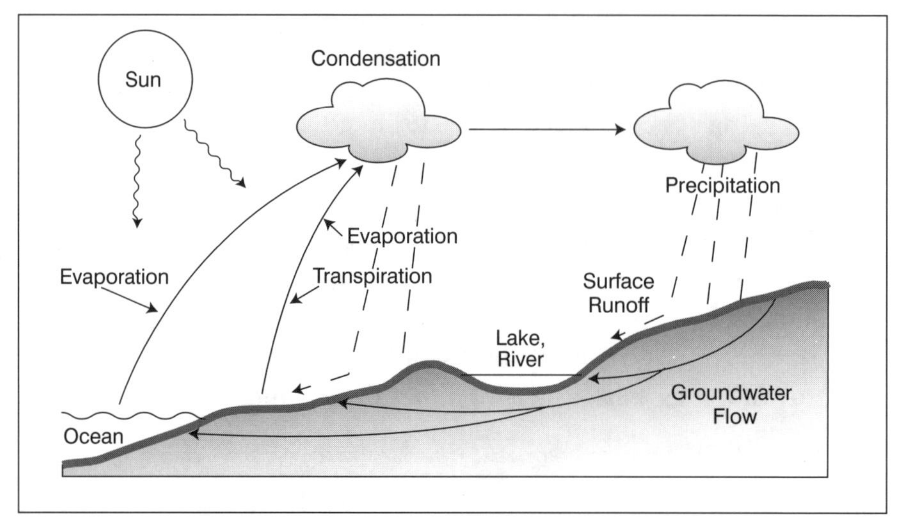

Figure 1. *Water cycle.*

1.5.2 Moore's Laws of Pollution Control and Hazardous Waste Management

1. *Population times per capita use of a resource equals total use of a resource. Also population times per capita generation of a waste equals total generation of a waste.*
   It is easy to forget that total use of a resource depends both on population and on per capita use of that resource. It is the *product* of the population and the per capita use that is important. This is illustrated in Figure 2, which shows the use of energy in this country since 1950, both as it has actually occurred and as it would have occurred had only the population increased (and not our per capita use of energy). It is clear that population growth and/or growth in per capita use of a resource cannot be sustained indefinitely. It is also true that population growth and/or growth in per capita production of waste cannot be sustained indefinitely. This will eventually take us outside one or another of the earth's cycles and lead to disaster.

2. *Don't let it get out in the first place.*
   This means that it is best to avoid releasing a pollutant or hazardous substance to the environment, so it won't have to be cleaned up later. Methods of avoiding release include:
   - changing the process so that a pollutant is not created,

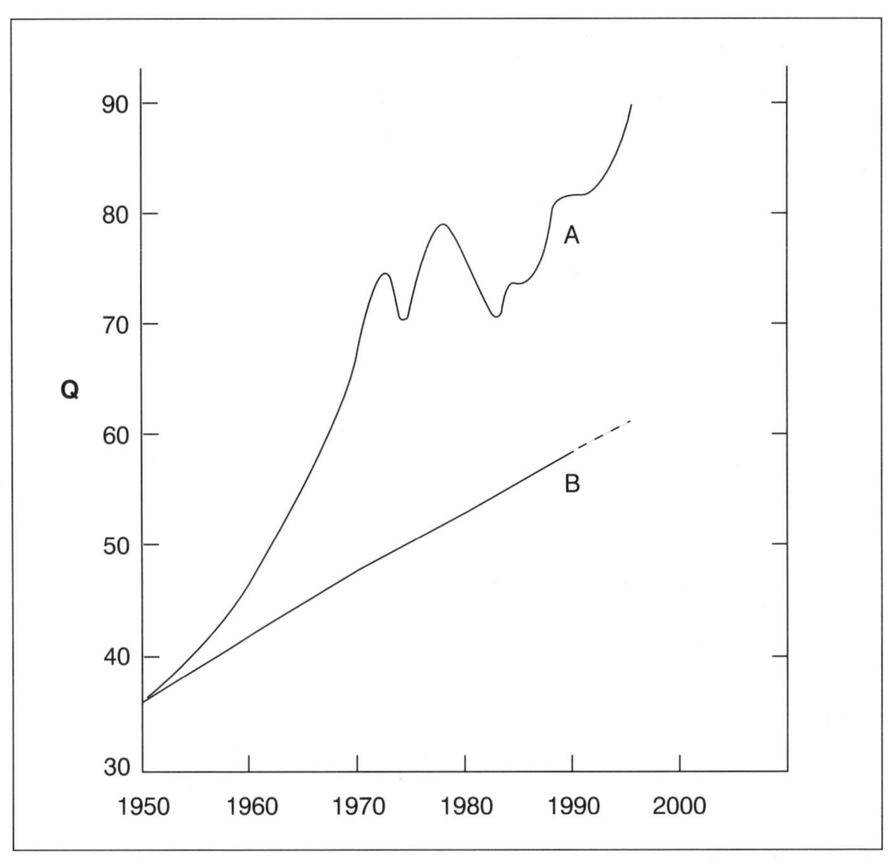

Figure 2. *Energy consumption in the United States. A) Actual consumption. B) Consumption with no increase in per capita use since 1950.*

- recycling the pollutant so that it is no longer hazardous, or
- providing controls so that the pollutant does not escape.

3. *Water and hazardous waste don't mix.*

   This is a variation on the chemistry student's old rule that glass and metal don't mix (i.e., metal ringstands in a drawer will often break glassware in the same drawer). Water and hazardous waste really do mix, sometimes very well. The result is that water moving downward through buried hazardous waste can dissolve or leach the hazardous constituents and carry them down through the vadose zone and into ground water, where the hazardous constituents can then move to wells, rivers, and on to plant, animal, and human consumers.

1.5.3 **Is Hazardous Waste Management a Science?**

Biology, toxicology, chemistry, and a bit of physics all contribute to hazardous waste management. Without toxicology, we would not know that some pollutants harm biota (biology). Without analytical chemistry, we would not know which pollutants harm biota or in what concentrations pollutants harm biota. Without physical chemistry, organic chemistry, inorganic chemistry, and physics, we would not understand the chemical and physical processes that occur or what is needed to manage the processes, i.e., the technology needed to manage or prevent pollution, to clean the environment, and to protect life. Thus, hazardous waste management is an interdisciplinary subject that includes many technologies which, in turn, are based on many fields of science.

1.6 **POLLUTION PROBLEMS; HAZARDOUS WASTE PROBLEMS**

Pollution control problems and hazardous waste management problems exist in the air, in surface water, in soils, and in ground water. Furthermore, problems in one medium become problems in other media, i.e., "all things are connected." A major hazardous waste management problem is hazardous waste buried in the ground (land disposal) that can leach into ground water. Also, the treatment of hazardous wastes can result in the emission of hazardous substances to the atmosphere and in the release of hazardous substances to surface and ground water. This further compounds hazardous waste management problems. The major land disposal, water effluent, and air emission pollution problems listed below are discussed in more detail in Chapter 3.

Land Disposal

- RCRA and CERCLA waste sites, polychlorinated biphenyls (PCBs), chlorinated dioxins, other chlorinated organic compounds, industrial spills, underground storage tanks (USTs), heavy metals
- Agricultural chemicals, including nitrates and pesticides, agricultural waste piles

- Deposition from air emissions, acid rain
- Lagoons, cribs, trenches, process ponds, sewage sludge
- Mine tailings piles
- Radionuclides
- Medical wastes.

Surface Water

- Sewage: bacteria, viruses, chlorinated organic compounds, phosphates, storm water
- Agricultural chemicals, irrigation return flows, nitrates, pesticides
- Acidification from acid rain
- Industrial discharges: chlorinated organic compounds, heavy metals, suspended solids, heat
- Oil spills
- Leachates from mine tailings piles: heavy metals, acids
- Radionuclides.

Ground Water

- Most of the above surface water problems
- Agricultural chemicals: nitrates, pesticides
- Leachate from waste sites
- Injection wells
- Radionuclides.

Air

- Automobile emissions: volatile organic compounds (VOCs), NO_x, lead, smog, CO, CO_2, particulates, O_3, peroxyacetylnitrate (PAN)
- Industrial emissions: incineration, heavy metals, VOCs, SO_2, CO_2

- Acid rain: H_2SO_4, HNO_3
- Global warming: CO_2
- Ozone depletion
- Indoor air pollution: tobacco smoke, cooking, radon, asbestos
- Radionuclides.

CHAPTER 1 REFERENCES

Ember, Lois. *Chemical and Engineering News*, p 11, May 27, 1985.

Ember, Lois. *Chemical and Engineering News*, p 20, June 19, 1989.

Environment Reporter, (ER) p 878, September 17, 1993.

Herrick, Carlyle. *Chemical and Engineering News*, p 2, February 21, 1994.

Hileman, Bette. *Chemical and Engineering News*, p 28, January 6, 1997.

CHAPTER 2

Chemistry

Chemistry is the study of the structure, behavior, and reactions of atoms and molecules in their three phases: solid, liquid, and gas. The traditional sub-disciplines of chemistry are inorganic chemistry, organic chemistry, physical chemistry, and analytical chemistry (Lewis 1993). Concepts, technologies, and results from these disciplines, as needed for material presented in this text, are reviewed below.

2.1 INORGANIC CHEMISTRY

Inorganic chemistry is the study of all of the elements of the periodic table except for the carbon (C) contained in the compounds of organic chemistry. Inorganic elements and compounds include metals, acids, bases, salts, ores, and radionuclides. For the purposes of hazardous waste management, the important components of inorganic chemistry are toxic heavy metals and radionuclides. Toxic heavy metals (according to Congress, see CWA Sec. 307(a)(1)) include antimony (Sb), arsenic (As), cadmium (Cd), chromium (Cr), copper (Cu), lead (Pb), mercury (Hg), nickel (Ni), silver (Ag), and zinc (Zn). Radionuclides include fission products, activation products, and naturally occurring radionuclides such as uranium and thorium. Radionuclides are discussed in Section 3.4.

2.2 ORGANIC CHEMISTRY

Organic chemistry is the study of carbon-containing compounds other than carbonates. Carbonates are considered part of inorganic chemistry. The carbon compounds of life, such as proteins, fats, and carbohydrates are organic compounds. So also are carbon compounds associated with life but not usually part of living systems, such as methane (CH_4) and petroleum products. Volatile organic compounds (VOCs) are small organic compounds (liquids) that evaporate easily at room temperature (or are gases at room temperature) and that contain carbon, hydrogen (H), and sometimes other elements such as oxygen (O). Chlorinated organic compounds are organic compounds that contain chlorine (Cl). These compounds include carbon tetrachloride (CCl_4), trichloroethylene (C_2HCl_3), tetrachloroethylene (C_2Cl_4), chlorinated dioxins, polychlorinated biphenyls, and many pesticides. Many chlorinated organic compounds are toxic and are thus of interest in hazardous waste management. (The term "halogen" includes "fluorine, chlorine, bromine, iodine, or astatine.")

Aliphatic organic compounds are organic compounds that consist of straight or branched chains such as ethane (C_2H_6), octane (C_8H_{18}), or isobutane (C_4H_{10}). Aromatic organic compounds are organic compounds that contain rings such as benzene (C_6H_6). Polycyclic aromatic hydrocarbons (PAHs) are organic compounds that contain two or more ring structures. Benzene is a common ring structure found in PAHs. Some PAHs are carcinogens.

Several important organic compounds that can become components of hazardous wastes are shown in Figure 3.

2.3 PHYSICAL CHEMISTRY

Physical chemistry is the study of the application of mathematics to chemistry, particularly the application of the mathematical laws of physics. Units of measure, solubility, density, specific gravity, acidity, temperature, pressure, vapor pressure, reaction kinetics, and thermodynamic equilibrium are discussed below.

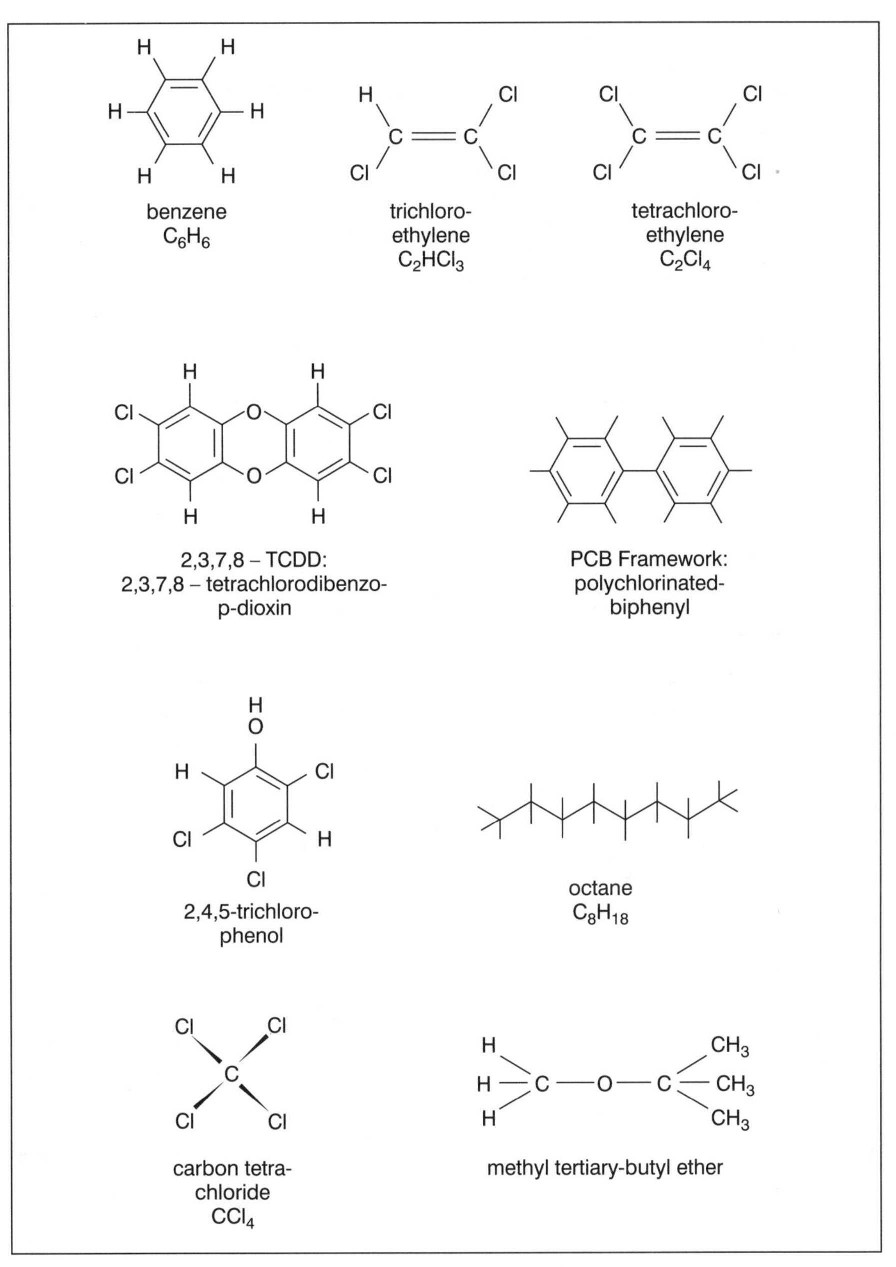

Figure 3. *Some chemical compounds of interest in hazardous waste management.*

2.3.1 **Units of Measure**

The established system of units of measure in scientific work today is the International System of Units, abbreviated SI (Nelson 1999). The basic units are the meter (length), kilogram (mass), second (time), mole (amount of substance), kelvin (thermodynamic temperature), ampere (electric current), and candela (luminous intensity). The symbols are m, kg, s, mol, K, A, and cd.

m	meter
kg	kilogram
s	second
mol	mole
K	kelvin
A	ampere
cd	candela

All other units are derived from these basic units. For the purpose of hazardous waste management, length, mass, time, amount of substance, and temperature are the most important. The older centimeter-gram-second (cgs) system of units is directly related to the SI system, and also in use today where the units are of more convenient size. Conversions needed for our use are:

$$100 \text{ cm} = 1 \text{ m}$$

$$1000 \text{ g} = 1 \text{ kg}$$

A unit of volume in frequent use, but whose size is not immediately obvious from the basic units, is the liter. One liter (1 L) is equal to 1000 cubic centimeters (1000 cm^3). Another such unit is temperature in degrees Celsius, sometimes called centigrade, (°C). For thermodynamic calculations temperature in kelvins (K) must be used. For laboratory measurements, temperature is often specified in degrees Celsius (°C). The conversion is:

$$t(°C) = T(K) - 273.15 \text{ K}.$$

The melting point of ice is 0°C or 273.15 K, and the boiling point of water is 100°C or 373.15 K. Physical constants in this book are from Lide 1994.

Because of differences of orders of magnitude in the quantities studied in chemistry, it is convenient to use prefixes that vary by factors of 10 to denote very small or very large quantities. For example, a centimeter is one hundredth of a meter, a millimeter is one thousandth of a meter, a kilometer is 1000 meters, etc. Some of the more common prefixes are as follows:

Prefix	Factor	Symbol
mega	10^6	M
kilo	10^3	k
centi	10^{-2}	c
milli	10^{-3}	m
micro	10^{-6}	μ
nano	10^{-9}	n
pico	10^{-12}	p

2.3.2 **Solubility, Density, and Specific Gravity**

Solubility is defined as the amount of substance dissolved in a given amount of liquid. For example, the solubility of calcium carbonate (calcite) in water at 25°C is 0.014 g/L and the solubility of sodium chloride (table salt) in water at 0°C is 357 g/L. Solubility is important in pump-and-treat technologies used for the removal of hazardous wastes from ground water. If the pollutant of interest is insoluble, then it will lie in a pocket above or at the bottom of the aquifer; and, unless the well casing directly contacts the pollutant, only water will be drawn to the surface for purification. The pollutant will be left below ground. Similarly, hazardous wastes that are only slightly soluble will be released to the ground water very slowly and will require very long pumping times for purification.

Density is the mass per unit volume of a liquid, solid, or a gas. For liquids and solids, density is usually measured in the laboratory in grams per milliliter (g/ml) or grams per cubic centimeter (g/cm^3). Water has a density of 1 g/cm^3 at 4°C. Specific gravity is the ratio of the density of a substance to that of water. For liquids that are insoluble in water, the density of the liquid determines whether the liquid will be found above water or below water when the liquid is trapped (as a pollutant) in an aquifer.

2.3.3 **Acidity (pH)**

The common measure of acidity in chemistry is pH. pH is the negative logarithm of the concentration (strictly speaking "activity") of hydrogen ions in water solutions. In water solutions, $[H^+][OH^-] = 10^{-14}$. This means that the concentration of hydrogen ions in moles/L times the concentration of hydroxyl ions in moles/L is equal to 10^{-14}. At neutrality, $[H^+] = [OH^-] = 10^{-7}$ and, $-\log [H^+] = 7$. Thus the pH of a neutral solution is 7. Acid solutions have $[H^+] > 10^{-7}$ and thus the pH is less than 7. In basic solutions, $[H^+] < 10^{-7}$ and thus the pH is greater than 7. Rainfall in equilibrium with carbon dioxide in the atmosphere has a pH of 5.5. Water percolating through municipal garbage dumps can have a low pH (i.e., be acidic) because of the formation of acetic acid from water and cardboard adhesives in the dump. The pH of vinegar and citric acid is about 2.5 and the pH of human blood is about 7.5.

2.3.4 **Temperature, Pressure, and Vapor Pressure**

As mentioned above, when temperature is used in any thermodynamic calculation, the temperature must be specified in kelvin (K). Otherwise, in scientific work, the temperature may be stated in degrees Celsius (°C).

Pressure is the force exerted per unit area on the wall of a container by a gas, liquid, or solid. Imagine a cylinder with a weightless piston inserted in the top of the cylinder to contain the substance inside. A gas will exert equal pressure on all sides of the cylinder, as well as upward on the bottom of the piston and downward on the bottom of the cylinder. This neglects gravitational effects, i.e., the cylinder is not a long one. A liquid will exert no pressure upward on the piston, an increasing amount of pressure on the side of the cylinder with increasing depth, and maximum pressure on the bottom of the cylinder. A rigid solid machined to the dimensions of the cylinder will exert no force upward on the bottom of the piston, no force on the side of the cylinder, and maximum force (pressure) on the bottom of the cylinder.

The unit of pressure in the SI system of units is the pascal (Pa), which is defined as the force exerted on a unit area measured in newtons (N) per square meter or N/m^2. One newton of force is one $kg{\cdot}m/s^2$. Thus the pascal has the units $kg/m{\cdot}s^2$. These units must be used in thermodynamic calculations in the SI system. For merely specifying pressure, however, millimeters of mercury (mm Hg) are frequently used. A standard air pressure (at sea level) of one atmosphere will support a column of mercury 760 mm high. Thus, pressures that are some fraction of one atmosphere are frequently expressed in mm Hg.

Vapor pressure means the pressure measured in the vapor (gas) phase in a sealed vessel containing a pure liquid with space above the liquid for evaporation. Vapor pressure is a measure of how easily a liquid will evaporate. Vapor pressure increases with temperature. Vapor extraction techniques for the removal of organic liquids from soil in the vadose zone depend upon the vapor pressure of the organic liquid. The vapor pressure of the organic liquid causes the liquid to evaporate into the flowing air used for extraction. The evaporated liquid, now a vapor, can then be condensed and removed from the air stream. A related technology is air stripping for the removal of an organic liquid from water. Air stripping will remove both water vapor and organic vapor because both the organic liquid and water have vapor pressures. Further separation of the two components is necessary after the vapor mixture is condensed. One method of separation depends on the difference in vapor pressure between the two liquids. If their vapor pressures are different, their relative concentrations may be different in the vapor phase than in the liquid phase at equilibrium. If this is true, then the two liquids can be separated by fractional distillation. The relationship between the phases (liquid and vapor) and the concentrations of each component in each phase must be determined experimentally (phase diagram). The vapor pressure of water at 25°C is 23.8 mm Hg and the vapor pressure of ethyl alcohol at 25°C is 59.0 mm Hg. Thus water and ethyl alcohol can be separated, at least partially, by fractional distillation. Whether or not full separation is possible depends on the phase diagram. A second method of separating an organic compound from water or water vapor is by adsorption of the organic compound on activated charcoal.

Solid inorganic compounds have very low vapor pressures and cannot be removed from the soil or water by vapor extraction or air stripping techniques.

2.3.5 Reaction Kinetics and Thermodynamic Equilibrium

Chemical reactions take time to occur. For example, it takes time for an organic liquid to evaporate into an air stream. Furthermore, chemical reactions take place in both directions. That is to say, at the same time an organic liquid is evaporating into an air stream, some of the organic vapor is condensing back into the liquid phase. If one waits long enough, thermodynamic equilibrium is established, which means that evaporation is taking place at the same rate as condensation and that the concentrations of the organic compound in the liquid and gas phases are not changing. A catalyst can hasten the attainment of equilibrium in some reactions. What this means is that when contemplating the use of a chemical reaction to treat a hazardous waste, attention must be paid to the kinetics of the reaction(s), to possible equilibria, and to possible catalysis, in order to optimize the process. Otherwise time, energy, and materials will be wasted.

2.4 ANALYTICAL CHEMISTRY

Analytical chemistry is the qualitative and quantitative branch of chemistry. Analytical chemistry is used to determine what the components of a sample are and how much of each component is present in the sample. Environmental science would not be science if it were not for the existence of analytical chemistry. We would not be able to identify hazardous substances or the amounts that are hazardous. Fortunately we can both identify and measure hazardous substances using analytical chemistry. Some techniques of analytical chemistry are discussed very briefly below.

Analytical measurements are based on some physical or chemical property of the atom or molecule in question. Before the development of modern instrumental analytical techniques, identification of an element or compound was usually based on a specific chemical reaction and on the measured masses or volumes of the reactants and the products. Sensitivities were not high. Now, most analytical measurements are based on instrumental methods where the atom or molecule (reactant and/or product) is probed or manipulated by an electric, magnetic, and/or gravitational field. Sensitivities are very high.

2.4.1 **Optical Spectroscopy**

The electromagnetic spectrum extends from very short wavelength gamma rays; through x rays; the ultraviolet, visible, and infrared regions of the spectrum; and on to long wavelength microwaves and radio waves (Figure 4). Much of the electromagnetic spectrum can be used for the purposes of analytical chemistry. The wavelength of the nominal visible spectrum extends from 400 nm (blue) to 800 nm (red), the ultraviolet spectrum includes wavelengths shorter than 400 nm, and the infrared spectrum includes wavelengths longer than 800 nm.

Visible and ultraviolet radiation induces characteristic electronic excitations in both atoms and molecules. These electronic excitations are caused by the absorption of radiation. This absorption of radiation can, in turn, be easily measured in a spectrometer. For example, atomic absorption spectroscopy may be used to analyze for metals after the metals are vaporized in a graphite furnace. Both visible and ultraviolet radiation can also induce characteristic fluorescence in atoms and molecules. That is, the absorption of radiation excites electrons to higher energy levels in atoms, after which the electrons relax to intermediate energy levels and emit fluorescent radiation. For example, atomic emission spectroscopy can be used to analyze metals after electrons in the metals are excited in an inductively

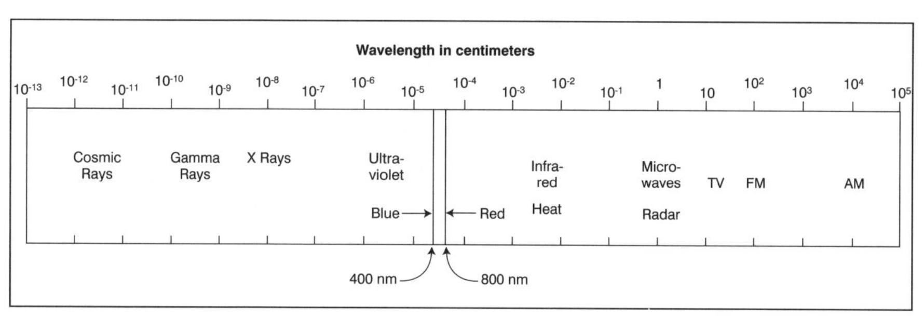

Figure 4. *Electromagnetic spectrum.*

coupled plasma. Infrared radiation induces characteristic vibrational transitions in organic molecules. Also, microwaves induce quantized rotational transitions. Again, the radiation that is absorbed during these transitions can be measured easily.

2.4.2 Electrophoresis

Electric fields can separate ions (charged atoms or molecules) in a gel by virtue of different rates of movement of the ions in the field. Some kind of calibration or detector must be used to identify the ions as they exit the electrophoresis apparatus.

2.4.3 Chromatography

Molecules (and ions) in the liquid or gaseous state may be separated from each other in a chemical matrix packed in a column by virtue of different rates of movement of the molecules through the chemical matrix. The different rates of movement are caused by different rates of adsorption and desorption of the molecules on the chemical substrate. Organic molecules, and also inorganic cations and anions such as chloride, nitrate, nitrite, calcium, sodium, may be separated by this technique. Calibration or further identification is required as the chemical species leaves the chromatograph, for example identification of the species with a mass spectrometer.

2.4.4 Mass Spectroscopy

Charged atoms and molecules (ions) move with different radii of curvature in a magnetic field, depending on both the mass and charge of the ion. The ratio of charge to mass can easily be determined, often leading directly to a positive identification of the ion.

2.4.5 Radioactivity

Radioactive nuclei decay with distinctive alpha, beta, and gamma energies. Gamma ray spectroscopy, for example, is a very well developed technique for identifying gamma emitters.

CHAPTER 2 REFERENCES

Lewis, Richard J. Sr. *Hawley's Condensed Chemical Dictionary*, 12th edition, van Nostrand Reinhold, 1993.

Lide, D. R., editor, *CRC Handbook of Chemistry and Physics*, 75th edition, CRC Press, Inc., 1994.

Nelson, Robert A. *Physics Today*, BG11, August 1999.

CHAPTER 3

Hazardous Substances and Hazardous Wastes

Important hazardous substances and hazardous wastes, their sources, and the environmental media in which they are found are discussed in Section 3.1. Environmental pathways for movement of pollutants between and among the land, water, and air are discussed in Section 3.2, the chemistry and toxicology of specific hazardous substances are discussed in Section 3.3, and radioactivity is discussed in Section 3.4.

3.1 POLLUTANTS AND HAZARDOUS SUBSTANCES

Pollutants, hazardous substances, and hazardous wastes are discussed here according to the medium in which they occur. However, since all things are connected, it should be obvious that cross pollution occurs between and among media.

3.1.1 Land Disposal

Land disposal (disposal in landfills) was a major treatment of choice for solid and liquid wastes until the passage of RCRA and CERCLA. These wastes range from benign municipal wastes to hazardous solid and liquid wastes. Wastes have been placed in the ground deliberately. They have also been deposited on the surface of the ground from the air or from rain, and have been spilled on the ground. Solid pollutants placed in dry soil tend to migrate slowly through the soil. On the other hand, liquid pollutants or solid pollutants in contact with liquids, especially water, can migrate rapidly through the soil. Hazardous wastes in soils can be problems because of direct human or other animal exposure, because of uptake by plants from the soil and subsequent consumption by animals and humans, and because of leaching into water and subsequent plant uptake and animal or human consumption. Small amounts of soil can be cleaned to reasonable levels, except that it is sometimes difficult to remove metals and some radionuclides.

RCRA and CERCLA Waste Sites, Industrial Spills, Sewage Sludge, Underground Storage Tanks

RCRA and CERCLA waste sites are those that are regulated directly by these two federal laws. There are also other waste sites that receive the attention of various state laws and regulatory agencies, as well as sites that have not yet received the attention of any regulatory agency. RCRA and CERCLA waste sites can include surface and subsurface sites contaminated by wastes such as heavy metals, for example Cr^{+6}; hazardous organic compounds, for example chlorinated aliphatic (chain) organic compounds such as trichloroethylene (TCE) and tetrachloroethylene, also called perchloroethylene (PCE); chlorinated aromatic (ring) organic compounds such as pentachlorophenol, chlorinated dioxins, and polychlorinated biphenyls (PCBs); insecticides, herbicides, and fungicides; industrial byproducts, including heavy metals and chlorinated organic compounds; sewage sludge that may contain bacteria and viruses; radionuclides; and leakage from underground storage tanks (USTs) including petroleum products (gasoline). Problems from waste sites in or on the ground arise from liquids in the wastes, or from precipitation, that leach the waste and carry it downward to ground water. Water plus the binding agent used in cardboard yields acetic acid in waste sites, which in turn can accelerate the leaching process. Also, anaerobic decomposition of organic compounds yields methane gas. Waste sites include municipal waste dumps that may have been legal at one time, as well as sites in the countryside where wastes were simply dumped a few barrels at a time.

Agricultural Chemicals in Soils, Nitrates

Agricultural pollutants in soils include nitrates from fertilizers, herbicides, insecticides, and nitrates from fertilizers. Many pesticides are chlorinated organic compounds.

Deposition from Air Emissions, Acid Rain

Depositions on soils from air emissions include heavy metals such as arsenic (As) and lead (Pb) from ore refining, and acid-formers such as NO_x, SO_2, and sulfates, often from coal-fired power plants.

Lagoons, Cribs, Trenches, and Process Ponds

Lagoons and process ponds have been used for the treatment of sewage and industrial wastes, particularly organic wastes. Cribs and trenches have been used in the past for disposal of radionuclides. Seepage of these wastes into the ground results in soil and groundwater contamination.

Mine Tailings Piles

Heavy metals such as arsenic (As), silver (Ag), cadmium (Cd), chromium (Cr), lead (Pb), and inorganic anions such as sulfates and phosphates occur in mine tailings piles. Drainage from mine tailings piles may lead to surface water acidification. For example, portions of Lake Coeur d'Alene, ID, have been contaminated with mine tailings washing down the Coeur d'Alene River from mining operations upstream.

Medical Wastes

Medical wastes include supplies (syringes, gloves, bandages) and infected materials which, if not treated before disposal, can pose a hazard to human health. Organic medical wastes are usually incinerated. Reusable medical supplies are disinfected or autoclaved.

Radionuclides

Radioactive substances emit alpha, beta, and gamma radiation, which can be harmful to both plants and animals. Radioactive substances occur naturally in uranium deposits, and also arise from man-made processes, such as electric power production from nuclear reactors and plutonium production for weapons purposes. Disposal of waste radionuclides from these processes has become a very substantial waste management problem. Releases to the atmosphere have occurred from weapons testing and from reactor processes. Releases to the soil and to bodies of water have also occurred from waste sites, waste tanks, and cribs. Heavy metal radionuclides are also toxic because of their chemical properties, in addition to their nuclear properties. See Sec. 3.4 for a more complete discussion.

3.1.2 Surface Water

Water, like the land, is (or was) a great receptor for pollutants, hazardous substances, and hazardous wastes. Fortunately, the passage of the Clean Water Act has decreased the use of bodies of water for disposal purposes in the United States.

Once a pollutant enters the water, it can become dispersed. If the pollutant enters the ocean, it can eventually become dispersed throughout the ocean, and thus impossible to clean up. On the other hand, if a pollutant enters a small lake or confined ground water, this water can sometimes be cleaned up. Small amounts of water can be cleaned to any desired level of purity, i.e., for drinking water purposes.

Sewage: Chlorinated Organic Compounds, Phosphates, Storm Water

Sewage and liquid industrial wastes discharged through sewage disposal plants can contain heavy metals, organic compounds including chlorinated organic compounds, bacteria, viruses, particulates (suspended solids), oil, and inorganic compounds such as phosphates. Chlorinated organic compounds can be formed from the reaction of the chlorine used to disinfect sewage with the organic compounds in sewage. Storm water can enter sanitary sewers during periods of high rainfall and cause overflows of raw sewage. Both the discharge of sewage and some storm waters to surface waters are now regulated under the Clean Water Act.

Agricultural Chemicals, Irrigation Return Flows

Agricultural surface water pollutants can include insecticides, herbicides, fungicides, and nitrates.

Acidification

Nitrous, nitric, sulfurous, and sulfuric acids from coal-fired power plants, automobiles, and industrial emissions (acid rain) can cause acidification of surface water. Chloride ion enters surface water as road salt or as sodium chloride (NaCl) from ocean spray, not usually as HCl. The pH of acid waters can be from 5.0 to 5.5. In the western U.S., surface and ground waters can be slightly basic. For example, the pH of the Columbia River at the Hanford Site in southeast Washington is approximately 8.0 (PNNL 1997).

Industrial Discharges: Chlorinated Organic Compounds, Heavy Metals, Suspended Solids, Heat

Direct industrial discharges to surface waters can include heavy metals such as Cd, Cr, and As from mining and plating industries; chlorinated organic compounds such as dioxins from the paper and herbicide industries; asbestos-like particulates (taconite tailings in Lake Superior) from mining; and radionuclides. Discharges of heat into bodies of water can affect fish, waterfowl, and aquatic organisms both adversely and beneficially. Many of these discharges are now regulated by the CWA.

Oil Spills

Oil spills can pollute bodies of water, shorelines, and land areas. In water areas, birds, fish, and aquatic mammals are especially at risk. Oil spills are regulated under the CWA and the Oil Pollution Act (OPA).

Leachates from Mine Tailings Piles

Leachates from mine tailings piles contain heavy metals and sometimes asbestos-like particles. These leachates have caused serious pollution problems in mining areas in Montana and Idaho and in rivers and lakes downstream.

3.1.3 Ground Water

Widespread ground water contamination is really not reversible, i.e., it is possible to clean a given amount of water to any desired level of purity in a batch or continuous process, but it is not possible to clean completely an aquifer (water remaining in the ground) because the water continues to leach impurities from the soil or to dissolve small amounts of slightly soluble liquids and/or solids. The original source of the contamination must be eliminated. Contaminants include those discussed above for surface water in particular, but also include those discussed below.

Agricultural Chemicals, Nitrates

Insecticides, herbicides, and nitrates are common ground water contaminants in agricultural areas. Forty-six different pesticides have been found in ground water in 26 states. Nitrate is even more common (OTA 1990).

Leachates from Waste Sites

Leachates from waste sites that contaminate ground water are the same as those mentioned above under *RCRA and CERCLA Waste Sites* to the extent that the wastes are soluble in water and/or acetic acid.

Injection Wells

Any liquid that can be put down an injection well can contaminate ground water, i.e., can contaminate the well next door or an entire aquifer.

3.1.4 **Air Pollution**

Air is also a receptor for pollutants, and air pollution is a problem. Usually, however, air pollution is not thought of as a hazardous waste management problem. Rather, air pollution is thought of as a pollution control problem. Hazardous wastes are not remediated by the cubic meter in air, as they are by the acre on land, or as they are by the gallon in water. Once a pollutant enters the atmosphere, it is widely dispersed, eventually throughout the entire atmosphere. This makes air, once it is widely polluted, impossible to clean. Although small amounts of air can be cleaned to any desired level of purity (in operating rooms for surgical purposes or in clean rooms for industrial purposes), only natural processes and natural cycles will clean the atmosphere as a whole. While some chemical alteration of pollutants can take place in the atmosphere, more often than not the natural cleaning process simply deposits the pollutant on the ground, where it may become a hazardous waste management problem if deposited in high enough concentrations.

Air pollution is regulated primarily under the Clean Air Act and not under RCRA or CERCLA. Nevertheless, some emission standards such as those for incinerators can be found in the RCRA regulations. Furthermore, some air pollution problems become water and land pollution problems and thus become RCRA or CERCLA problems (radioactivity, for example) and, finally, some air pollution control technologies are central to some hazardous waste cleanup technologies, such as incineration. Therefore, air pollution and air pollution control are discussed in this text.

Automobile Emissions: VOCs, NO_x, Lead, Smog, CO, CO_2, Particulates O_3, PAN

Automobile emissions include volatile organic compounds (VOCs) and oxides of nitrogen, which can combine under the influence of sunlight (photochemical oxidation) to create ozone and peroxyacetylnitrate, $CH_3COOONO_2$ (PAN), which are major constituents of smog. Other automobile emissions include carbon monoxide from incomplete combustion of gasoline and, of course, carbon dioxide and water vapor. The phaseout of leaded gasoline has effectively removed tetraethyl lead as an automobile combustion product.

Carbon dioxide stays in the air or is dissolved in the oceans, and oxides of nitrogen remain in the atmosphere or fall to earth as acid rain. Diesel exhaust frequently contains substantial quantities of carbon particulates. Under the Clean Air Act, National Ambient Air Quality Standards exist for six so-called criteria pollutants: particulates, ozone, sulfur dioxide, oxides of nitrogen, carbon monoxide, and lead. Under the Clean Air Act Amendments (CAAA) of 1990, Congress directed the U.S. Environmental Protection Agency (EPA) to evaluate 189 substances for potential regulation as hazardous air pollutants.

Industrial Emissions: Hazardous Emissions, SO_2

Industrial emissions include particulates (carbon soot and dust) and aerosols from incinerators and other sources, and sulfur dioxide and sulfate particles from coal-fired power plants. Sulfate particles are major contributors to reduced atmospheric visibility. Many trace elements can be products of combustion, and are often emitted as constituents of fly ash (particulates). These trace elements include heavy metals such as As, Hg, and Pb from smelters, coal-fired power plants, and incinerators. Other industrial emissions include VOCs such as organic compounds from refineries, oxides of nitrogen (to some extent) from combustion processes, carbon dioxide and chlorinated dioxins from combustion processes, and radionuclides. Many of these emissions are regulated under the Clean Air Act.

Acid Rain

Oxides of sulfur and oxides of nitrogen plus water create nitrous acid, nitric acid, sulfurous acid, and sulfuric acid. The oxides of sulfur are from industrial processes and coal-fired power plants, and the oxides of nitrogen are also from combustion processes, frequently from automobile engines. Carbon dioxide (CO_2) from combustion processes plus water (H_2O) in the atmosphere yield carbonic acid (H_2CO_3). Rainfall with a pH of 5.5 results from the equilibrium of CO_2 in air with rain droplets to form carbonic acid. The pH of acid rain (containing oxides of sulfur and nitrogen) can be between 3 and 5. Values as low as 2 have been observed. For comparison, the pH of vinegar (acetic acid) and lemon juice (citric acid) is about 2.5. Human blood is slightly basic with a pH of 7.5.

Global Warming

Emissions of carbon dioxide, methane, nitrous oxide, water vapor, and chlorofluorocarbons to the atmosphere contribute to global warming and are sometimes called greenhouse gases. These compounds transmit visible and ultraviolet radiation from the sun, but absorb infrared radiation that is radiated from the earth, thereby trapping heat in the earth's atmosphere. The global increase in carbon dioxide is mainly from the combustion of fossil fuel. The concentration of carbon dioxide in the atmosphere has been measured continuously since 1957 on Mauna Loa by Charles Keeling (and by others at other places more recently). The atmospheric concentration shows annual variations of about 6 ppm and has risen from about 316 ppm in 1957 to about 360 ppm in 1997. Carbon dioxide emitted from combustion of organic compounds remains in the air or is dissolved in the oceans. Some scientists estimate that a doubling of the concentration of carbon dioxide in the earth's atmosphere would increase the temperature at the earth's surface by 1 to 4°C. Conversely, removing all of the carbon dioxide from the earth's atmosphere would reduce the earth's temperature by approximately the same amount. Global warming could lead to partial melting of the polar ice caps and a corresponding rise in sea level, shifts in the locations of important agricultural and ecological regions, shifts in the location of deserts, etc. (See Karl et al. 1997 and references therein).

Methane gas arises from various sources, including the anaerobic decay of organic matter in garbage dumps and in sewage sludge.

Water vapor is the major greenhouse gas, which keeps the earth's temperature at +15°C rather than –18°C. Water vapor may lead to positive feedback in global warming, i.e., a little warming will cause more water to evaporate from the oceans, thus increasing the amount of water vapor in the atmosphere, thereby increasing global warming. On the other hand, water vapor produces clouds, and aerosols such as sulfates from coal-fired power plants scatter and reflect short wave-length sunlight and provide seeding for clouds, which reflect sunlight and thereby reduce global warming (Charlson et al. 1992). Some scientists suggest that global warming was less in 1992 than in 1990 and 1991 due to SO_2 from the eruption of Mount Pinatubo.

Emissions of carbon dioxide are not regulated (automobiles and coal-fired power plants are the main sources). Should they become regulated in the future, this regulation may have an impact on the incineration of organic compounds as a hazardous waste management technology.

Ozone Depletion

Chlorofluorocarbon (CFC) compounds used as refrigerants, such as trichlorofluoromethane and 1,1,2-trichloro-1,2,2-trifluoroethane, and other non-fluorinated compounds such as 1,1,1-trichloroethane and carbon tetrachloride, are stable in the lower atmosphere when released at the surface of the earth. When they rise to the stratosphere, however, chlorine atoms, which are free radicals, are stripped from the molecules by ultraviolet light. The chlorine radicals then catalyze the decomposition of ozone to oxygen in the upper atmosphere (Molina and Rowland 1974, Hamill and Toon 1991). This increases the amount of the sun's ultraviolet (UV) radiation that reaches the earth's surface. Ozone absorbs radiation of wavelengths shorter than 320 nm, and oxygen absorbs radiation shorter than 240 nm. Ozone depletion can thus increase the transmission of UV radiation to the earth between 240 and 320 nm. The absorption spectrum of DNA occurs at wavelengths shorter than 320 nm, so ozone depletion can increase the absorption of UV radiation by DNA, causing changes in DNA that lead to skin cancer. UV also causes cataracts. Almost all of the ozone in the lower stratosphere over Antarctica disappeared for 6 weeks in September and October (spring) 1987, 1989, and 1990 (Zurer 1991). Ozone depletion was first observed over the Antarctic, then over the Arctic, and is now observed worldwide. Also, bromine from methyl bromide and from halons (fluorocarbons that contain bromine) will cause the same effect. Direct observation of the increase of UV-B radiation at the earth's surface is not easy because of the interference of clouds and air pollution. Nevertheless, careful measurements are confirming that there is an increased transmission of UV-B to the earth's surface. (UV-B is radiation from 280 to 320 nm; UV-A is radiation from 320 to 400 nm.)

Indoor Air Pollution: Tobacco Smoke, Cooking, Radon, Asbestos

Tobacco smoke, formaldehyde and asbestos from some construction materials, and radon gas from soils and concrete are indoor pollutants. Nitrosamines are formed during the cooking of meat cured with nitrites and may be carcinogenic, i.e., N-nitrosodimethylamine ($(CH_3)_2NN{=}O$). The nitrite itself, of course, may cause methemoglobinemia, which is particularly harmful in infants. Asbestos, used in pipe insulation and in other construction materials, is regulated under three federal laws: the CAA, the TSCA, and the Occupational Safety and Health Act (OSHA). National Emission Standards for Hazardous Air Pollutants (NESHAP) under the CAA provide regulations for reporting requirements and for the application, use in manufacturing, removal and demolition, and disposal of asbestos; TSCA provides for more reporting requirements and for the removal of asbestos from schools; and OSHA regulates occupational exposures. In 1991, an EPA ban on some uses of asbestos was overturned by the Fifth Circuit Court of Appeals.

3.1.5 Summary

While land and water are the two environmental media of greatest interest with respect to hazardous waste management, air is also of interest if only because hazardous waste treatment technologies such as incineration involve releases to the atmosphere. All three media are, of course, of interest with respect to pollution. Pollutants are, or have been, released to all three media and can migrate from one medium to another, i.e., "All things are connected." Federal and state regulations apply to pollutants and hazardous wastes in all three media. For example, a long list of regulated hazardous substances in all three media appears in 40 CFR 302. Pollutants and hazardous wastes of interest can be grouped as follows:

Heavy (toxic) metals: As, Pb, Cd, Hg, Se, Ag, Cr, (Be)

Organic compounds: VOCs, benzene, chlorinated organic compounds (pesticides, dioxins, PCBs, CFCs, etc.)

Oxidants: VOCs + NO_x + sunlight = photochemical oxidants (ozone and PAN)

Acid formers: SO_2, NO_x, CO_2

Bacteria, viruses

Radionuclides

Particulates: asbestos, carbon soot, suspended solids

Heat.

3.2 ENVIRONMENTAL PATHWAYS

Once a pollutant enters the environment, it can reach humans and other biota by a variety of pathways. Air pollutants can reach plants and animals directly, or can be deposited on water or on soils and crops, and can reach animals through crops. Water pollutants can reach plants and animals directly from rainfall or irrigation, or can reach animals through food. Pollutants discharged to the soil may stay in place if dry, but if wet will eventually reach ground or surface water and then will reach plants and animals through the water pathway. A good example is the path sometimes taken by the radioactive species I-129 which may be released to the atmosphere from nuclear activities. I-129 is carried to the ground and vegetation by rainfall, is eaten by cows along with the vegetation, is incorporated into the cow's milk, and is in turn ingested by humans. The I-129 is then taken up by the human thyroid gland where radiation from I-129 can cause thyroid cancer. See Figure 5.

3.3 CHEMISTRY AND TOXICOLOGY

The word "toxic" means poisonous. Human toxicology covers 1) short-term acute and long-term chronic effects on the eyes, nose, lungs, skin, nervous system, and internal organs; 2) longer-term carcinogenic (cancer causing) effects; 3) teratogenic effects (birth defects); and 4) mutagenic effects (inherited defects). Toxic substances can affect tissues, cells, and molecules. For example, toxic substances can irritate the lungs (tissue), can interfere with membrane processes (cell), or can disrupt DNA (molecule).

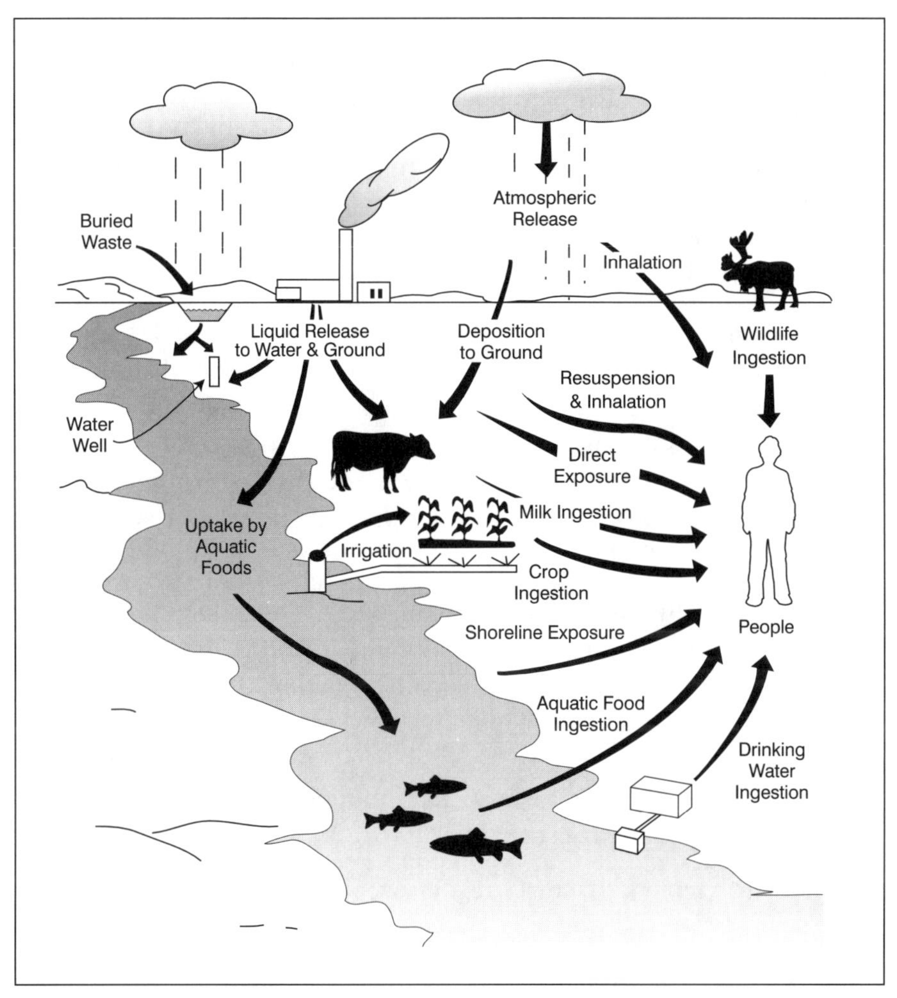

Figure 5. *Environmental pathways.*

In this section, the toxicology and chemistry of hazardous wastes, as well as the toxicology of some air pollutants related to hazardous waste treatment, are discussed briefly. Toxicological profiles have been prepared for a number of hazardous substances by the Agency for Toxic Substances and Disease Registry in Atlanta, GA. These profiles may be consulted for more information. (See also Zakrzewski 1991.) Chemical constants in this section and other sections of the book are taken from Lide 1994. Most of the substances discussed in this section are regulated under one or more environmental laws, some of which are shown in parentheses. OSHA also regulates many of these substances in the work place.

The toxicity of these substances is important because of their effects on human health and the environment. The physical and chemical properties of these substances are important because they dictate the ease or difficulty of removing them from the environment.

- **Acid rain, sulfuric and nitric acids, sulfates.** These are eye and lung irritants and can dissolve limestone (statues and buildings). NO_2 can kill lung cells, and can cause methemoglobinemia. NO_2 dimerizes to N_2O_4, which is a very corrosive liquid. Also acid rain leads to fish kills and damage to trees and plants through leaching of nutrient cations from the root zone (Likens et al. 1996). Acidic water at a pH of 6.0 to 6.5 can give rise to biological effects. (CAA)
- **Asbestos (a magnesium silicate).** Lung cancer, asbestosis (scarring of the lungs), and mesothelioma (cancer of the lining of the chest or abdomen) in asbestos workers can result from inhalation of asbestos fibers. Asbestos stimulates growth of collagen fiber. Ninety-five percent of the asbestos used in the U.S. is chrysotile, which is the least toxic form. Disagreement exists about hazards of non-occupational exposure. (CAA, TSCA)
- **Bacteria, viruses.** Bacteria and viruses can cause disease.
- **Benzene.** Benzene damages blood cells and causes leukemia, i.e., benzene is a carcinogen (MP 5.5°C, BP 80.0°C, specific gravity 0.88 at 20°C, slightly soluble in water). (CAA, RCRA)
- **Beryllium (Be).** Inhalation of Be dust causes lung scarring, which is sometimes fatal. Be is a suspected carcinogen. (CAA)
- **BTEX.** BTEX is an acronym for benzene, toluene, ethyl benzene, and xylenes. Toluene, ethyl benzene, and xylenes are derived from benzene. All four are volatile, toxic, and constituents of hydrocarbon fuels.
- **Chlorinated organic compounds.** See "Halogenated organic compounds."

- **Chlorofluorocarbons (CFCs).** Chlorofluorocarbons (CCl_2F_2, $CFCl_3$, and CCl_2FCClF_2, for example) are used as refrigerants, have a 100-year lifetime in the atmosphere, and are responsible for ozone loss in the stratosphere. CFCs also burn to form phosgene, $COCl_2$, which hydrolyzes to HCl and CO_2 in the lungs. HCl causes severe edema. Phosgene was used as a poison gas in World War I. (CAA)

- **CO.** Carbon monoxide gas reacts with the iron in hemoglobin more strongly than oxygen and carbon dioxide, and thus prevents oxygen and carbon dioxide from being transported within the body. Suffocation may result. (CAA)

- **CO_2.** Carbon dioxide gas is heavier than oxygen, may accumulate in low places (wells, for example), and displace oxygen in the air that is needed for breathing. In August 1986 at Lake Nyos in Cameroon, West Africa, 1746 people were killed by the release of carbon dioxide that was dissolved deep in the cold volcanic-lake water. The carbon dioxide came out of solution, as in an opened carbonated beverage bottle, spread over lakeshore lowlands, flowed down the slope, and killed people and animals in villages below the level of the lake (Ladbury 1996).

- **Dioxins.** Chlorinated dioxins consist of two benzene rings linked by two oxygen atoms, with varying numbers of chlorine atoms replacing hydrogen atoms on the benzene rings. Seventy-five chlorinated dioxins exist, of which 2,3,7,8,-tetrachlorodibenzo-p-dioxin (TCDD) is the most notorious. Dioxins are byproducts in the manufacture of 2,4,5-trichlorophenol, which is used in the synthesis of herbicides, and are contaminants in the herbicide agent orange (2,4,5-T and 2,4-D) (Tschirley 1986). They are produced in chlorine bleaching of pulp in the paper industry and in incomplete burning of municipal waste. Ninety-five per cent of dioxin emissions in the United States are estimated to arise from the incineration of medical and municipal wastes (ES&T 1994). Chlorinated dioxins are related to and an impurity in pentachlorophenol, which is a wood preservative. Chlorinated dioxins are not and never have been produced for commercial purposes. TCDD has a melting point of 307°C, is practically insoluble in water, and has an LD_{50} of 1 μg/kg when ingested in guinea pigs and 5000 μg/kg in hamsters. (LD_{50} means a lethal dose to 50 percent of the test animals.) Chloracne is the worst effect in humans, although diabetes and temporary effects on the liver and nervous system have been observed. No deaths were caused at Seveso, Italy, when 2 kg were released in 1976. However, higher than normal rates of some types of cancer are now beginning to be observed, although there is not yet an overall increase in cancer. Dioxins are not yet known to be potent human carcinogens, although they may be human carcinogens in high doses (Abelson 1983, Tschirley 1986, EPA 1997). EPA, as of September 1994, lists dioxin as a probable human carcinogen (ES&T 1994). Low doses can cause endometriosis in female monkeys and may be related to endometriosis in women. Natural degradation of dioxin occurs in sunlight. Little degradation occurs in soil. Thermal destruction requires a temperature of 700°C. (TSCA)

- **Gasoline.** Gasoline is a mixture of organic compounds and additives in which branched octanes are the nominal components. Straight chain octane has a melting point of -56.8°C, a boiling point of 125.6°C, a specific gravity of 0.70 at 25°C, and is insoluble in water. Gasoline may contain small amounts of BTEX and formerly contained tetraethyl lead, which can cause lead poisoning.

- **Halogenated organic compounds.** (A halogen atom is a fluorine, chlorine, bromine, or iodine atom.) Trichloroethylene (MP -84.7°C, BP 87.2°C, slightly soluble in water, specific gravity of 1.46 at 20°C) is used for degreasing; occurs at many Superfund sites; is fairly easily reduced; is subject to bioremediation in concentrations up to 100 ppm; causes eye, nose, and throat irritation; and can cause damage to the nervous system. Perchloroethylene, also called tetrachloroethylene, (MP -22.3°C, BP 121°C, insoluble in water, specific gravity 1.62 at 20°C) is used in dry cleaning and causes the same effects as trichloroethylene. Carbon tetrachloride is a solvent (MP -23°C, BP 76.8°C, very slightly soluble in hot water, specific gravity 1.59 at 20°C), is used in the synthesis of fluorocarbons, and causes the same effects as trichloroethylene. All three cause liver and kidney damage and are potential human carcinogens. Halogenated organic compounds also are emitted to the atmosphere by forest burning (CH_3Cl) and from live spruce trees. Cl_2 (from water disinfection) and organic matter in sewage can react to form halogenated organic compounds (disinfection byproducts) in the treated water. (CAA, RCRA)

- **Heavy metals.** Heavy metals are taken into the body by inhalation and ingestion. They can cause neurological abnormalities. As and Cd dust from mining operations and smelters can cause lung cancer. As dust can cause skin lesions, Zn dust can cause anemia, and Cd dust can cause kidney disease. Pb can also damage bone marrow. Cr causes kidney damage and lung cancer. Heavy metals may interfere with nutrient-cycling processes in the root zone of trees and plants by inhibiting microbiological processes in soil. (CAA, RCRA)

- **Mercury.** Mercury is a liquid metal with a relatively low vapor pressure compared to many organic liquids, but high enough to provide toxic amounts of mercury in closed rooms. It was used as a catalyst in a chemical process and was discharged in a sludge into Minamata Bay, Japan. Aquatic biota converted the mercury to methyl mercury, which accumulated in fish and shellfish and caused methyl mercury poisoning in persons who ate the fish and shellfish (Zakrzewski 1991). Damage includes kidney, bone, brain, and nervous system damage. Mercury is also emitted to the atmosphere from municipal and medical incinerators and from coal-fired power plants (EPA 1996). (CAA, RCRA)

- **MTBE.** MTBE is an acronym for methyl tertiary-butyl ether. MTBE is used as a gasoline additive to provide extra oxygen to oxidize the combustion product CO more completely to CO_2. Gasoline additives were mandated by the Clean Air Act Amendments of 1990 for areas that exceed the National Ambient Air Quality Standards for CO in the winter. Unfortunately, MTBE is toxic, is soluble in water, is volatile, does not biodegrade easily, is listed as a hazardous air pollutant under Sec. 112 of the CAA, and may be a carcinogen. EPA has not yet set a drinking water standard for MTBE (Cooney 1997), although drinking water wells in Santa Monica have been shut down because of MTBE contamination. Ethyl alcohol is also used as a gasoline additive to provide extra oxygen. (CAAA)

- **Nitrates.** Nitrates are reduced to nitrites in the body, which then react with the iron in hemoglobin, i.e., the nitrite displaces the oxygen and carbon dioxide that normally attach to the iron in hemoglobin. This can cause methemoglobinemia, particularly in infants.

- **Nitrosamines.** Nitrosamines are carcinogens. They are formed from nitrites (used to cure meat) and amino acids (from proteins in meat).

- **Ozone, O_3.** Ozone gas is a strong oxidizing agent (a photochemical oxidant, which means an oxidant produced photochemically) that causes chronic and acute pulmonary effects, and possibly accelerated aging of the lungs. Ozone can cause rapid damage to crops and vegetation (decline in growth or yield) by interfering with carbon metabolism (respiration and photosynthesis), and may destroy chlorophyll. Ozone (good ozone) is formed in the stratosphere by the photolysis of oxygen at wavelengths shorter than 242 nm. Ozone (bad ozone) is formed in the troposphere by the photocatalyzed reaction of NO_2 and VOCs at wavelengths less than 400 nm. Ozone absorbs radiation at wavelengths less than 320 nm, thereby reducing ultraviolet flux at the earth's surface. (Oxygen absorbs at wavelengths less than 240 nm.) (CAA)

- **PAH.** PAH is an acronym for polycyclic aromatic hydrocarbon. PAHs are organic compounds containing two or more ring structures. Benzene is the most common ring structure found in PAHs.

- **Particulates.** Particles less than 1 micron in diameter can be trapped in the lungs and cause lung disease (black lung in coal miners is caused by SiO_2, not C) and lung cancer in uranium miners (from radon decaying to other radioactive chemical species either in the lung or before inhalation). Sulfate particulates cause reduced atmospheric visibility. The 24-hour national ambient air quality standard for particles 10 microns or less in diameter is 150 micrograms per cubic meter for one day per year. For particles 2.5 microns or less in diameter, the proposed standard is 65 micrograms per cubic meter. In peacetime London in the 1940s and 1950s, concentrations rose to 500 micrograms per cubic meter. Sulfate particulates in the atmosphere may reflect UV and nucleate clouds that also reflect UV, thereby cooling the earth and compensating for global warming. Smaller particulates are from man-made combustion sources such as coal-fired power plants and internal combustion engines. Larger particulates are from natural sources such as soil dust. (CAA)

- **Peroxyacetylnitrate (PAN), $CH_3COOONO_2$.** PAN is a component of smog that causes lung and eye irritation. (CAA)
- **Pesticides: herbicides, insecticides, and fungicides.** These are poisons. But not all are human poisons. For example, DDT (2,2-bis(p-chlorophenyl)-1,1,1-trichloroethane), an insecticide, is remarkably nontoxic to mammals, has not been shown to cause harm to humans, and has saved thousands of lives. However, it is highly soluble in lipids, undergoes bioconcentration in the food chain, concentrates in birds, interferes with calcium metabolism, and results in thinning of the egg shell in some species and in the loss of chicks (for example, the peregrine falcon). [Federal Insecticide, Fungicide and Rodenticide Act (FIFRA)]
- **Polychlorinated biphenyls (PCBs).** PCBs consist of two benzene rings joined by a single bond with varying numbers of chlorine atoms replacing the hydrogen atoms on the benzene rings ($C_{12}H_{10} + Cl_2$). PCBs were once used as insulating fluids in high voltage transformers. They have low vapor pressures, are soluble in lipids, are persistent in the environment, and have long elimination times from humans. PCBs cause liver damage, and are potential carcinogens. They cause liver cancer in rats at high chlorine content, but less liver cancer than in controls at low chlorine content. Manufacture of PCBs ceased in 1978 (Abelson 1991, EPA 1997). (TSCA)
- **Radioactivity** (see Sec. 3.4 below). Examples of radionuclides include I-129, I-131, Sr-90, Co-60, Cs-137, and radon. Radiation causes cell damage, DNA mutations, and cancer. I-129 and I-131 are absorbed by the thyroid gland and may cause thyroid cancer. Sr-90 is a bone seeker (replaces calcium) and can cause various kinds of cancer. Cs-137 goes to the muscles. I-129, I-131, and Sr-90 from fallout from above-ground atomic weapons testing get into milk. (Above-ground radioactive weapons testing is now essentially non-existent.) Co-60 is a hard gamma emitter that can cause leukemia and other kinds of cancer. Radiation can kill cells outright or alter DNA and lead to cancer and birth defects. Radon is a gas which can be inhaled. If it decays while in the lungs, its

radioactive daughters may not be expelled. The later alpha emissions are very destructive to adjacent cells. There are several different isotopes of radon, which result from the decay of isotopes of uranium and thorium. (CAA, SDWA, AEA)

- **Smog.** Smog components such as ozone and PAN cause eye and lung irritation. (CAA)
- **Volatile organic compounds (VOCs).** VOCs are small molecular weight organic compounds that are gases at room temperature or are liquids with high vapor pressures that easily evaporate into the atmosphere. They react with NO_x to form ozone and PAN. Refinery operations and incomplete combustion of organic compounds are major sources of VOCs. Trees are also sources of VOCs. (CAA)
- **Summary.** Toxicological effects of hazardous substances on humans include;

  Irritation (skin/eyes/lungs)

  Asphyxiation

  Lung damage

  Kidney damage

  Liver damage

  Blood component damage

  Neurological effects

  Birth defects

  Mutations

  Cancer.

3.4 **RADIOACTIVITY**

Because of its importance, radioactivity deserves a section of its own. Radioactivity, i.e., radioactive waste, cannot be treated or destroyed by biodegradation or incineration. The only treatments are shielding, distance, and time. Radioactivity is the spontaneous decay or conversion of an unstable atomic nucleus into another atomic nucleus (or nuclei) accompanied by the emission of one or more kinds radiation, including heat. The kinds of radiation of particular interest from a hazardous waste management point of view are alpha particles, beta particles, and gamma rays. Alpha particles are helium nuclei, beta particles are electrons, and gamma rays are high energy x-rays. All radiation shows the physical properties of both waves and particles, but alpha and beta particles are usually considered to be particles, while gamma rays are usually considered to be waves. Another particle of interest around reactors is the neutron, which is a particle with the mass of a proton but with no charge. Neutrons are emitted when a nucleus such as U-235 fissions into two unequal-sized nuclei approximately half the mass of the parent U-235.

Uranium occurs naturally and consists mostly of U-238 with approximately 0.7 percent U-235. Uranium is enriched to 2 or 3 percent U-235 for use in power reactors, and is enriched more highly for use in plutonium-production reactors.

In a reactor process, the U-235 atom fissions releasing two fission-product atoms, neutrons, and heat. Neutrons are absorbed by U-238 (always present in reactor fuel in larger amounts than U-235) which, in a short series of nuclear reactions transmutes into Pu-239. Both U-235 and Pu-239 are fissionable, i.e., both can spontaneously or by induction decay into two roughly equal nuclei, and both can be used in fission reactors. In a fission reactor, heat can be used to generate steam, which can then be used to power a turbine connected to a generator for the purpose of producing electricity. Plutonium can be extracted from spent fuel in a reprocessing facility and used for weapons purposes or to power a plutonium-fueled fission reactor for the purpose of generating more electricity. The plutonium production process makes use of the U-238, which is in much greater supply than U-235. This is the principle of the breeder reactor, i.e., neutrons from the fissioning of U-235 react with U-238 to produce Pu-239. Thus, fissionable Pu-239 is produced from non-fissionable U-238. In a reactor, other radioactive species are also produced by fission, i.e., fission products such as Sr-90 and Cs-137. Also, activation products are produced, such as Co-60. Activation is the absorption of a neutron to create a new radioactive species such as Co-60. Co-60 is produced by the absorption of a neutron in Co-59, a naturally occurring, nonradioactive isotope in steel.

Radioactivity is described by the number of disintegrations per second occurring in a given amount of the radionuclide and by the half life, which is the time for one-half of a given amount of the radionuclide to decay. A curie (Ci) is $3.7 \times 10^{+10}$ disintegrations per second. One becquerel (Bq) is one disintegration per second. Doses of radiation to individuals are described in rems. For comparison purposes, background radiation in central Washington is approximately 300 millirems per year. About 200 millirems come from radon and radon daughters, while approximately 100 millirems come from other sources such as cosmic rays and K-40. The dose to a population is described in person-rems, i.e., the dose received by each individual summed over the population.

Physiologically, alpha particles will not penetrate the skin, but once an alpha emitter is taken into the body, by inhalation for example, alpha particles are very damaging to tissues surrounding the alpha emitter. Beta particles can penetrate the body to a depth of a centimeter or so, and thus give rise to an external or internal dose depending on whether or not the beta emitter is inside or outside the body. Gamma rays are very energetic and will pass entirely through the body, while creating damage along the way. All three forms of radiation damage human tissue by ionizing atoms and molecules in the body, including DNA, and can cause cancer.

From a hazardous waste management point of view, gamma emitters such as Co-60 and Cs-137 (actually Ba-137m) are of concern as hazards to workers handling the waste, while other isotopes such as C-14, Cl-36, Sr-90, Tc-99, I-129, I-131, Pu-239, and U-235 and U-238 and their daughters are of concern as hazards to the public from short-term releases from nuclear facilities and from long-term migration from radioactive waste sites.

Radioactive wastes are classified as spent nuclear fuel, high-level waste, low-level waste, and transuranic waste. Spent nuclear fuel is nuclear fuel that has been irradiated in a reactor and contains U-235, U-238, various isotopes of plutonium, and fission products. Spent nuclear fuel is very radioactive and must be handled under water (water is a good radiation shield) or in heavily shielded casks and/or facilities usually made of lead and/or concrete. High-level waste is initially a liquid and is the residual material after plutonium has been removed from spent nuclear fuel. Often uranium isotopes and other isotopes such as Sr-90 and Cs-137 are also removed from spent nuclear fuel. High-level waste is also very radioactive and must be handled in shielded facilities. Long-term, but supposedly temporary, storage of government-owned high-level waste is in steel underground tanks. These tanks have leaked at both the Hanford and Savannah River Sites. The long-range plan is to vitrify these wastes and store them in a not-yet-constructed repository at Yucca Mountain, NV. Low-level waste exhibits a range of radioactivity and consists of contaminated materials, activated materials, spent solutions, etc. Some low-level waste may be handled in wooden crates, other low-level waste must be handled in secure containers and casks. Transuranic waste contains isotopes with an atomic number greater than 92 (92 is the atomic number of uranium). Some transuranic wastes may be contact-handled once they are in a secure container, others must be remotely handled. Mixed waste is waste that contains both radioactive and hazardous wastes. It must be managed in a way that accounts for both components.

Commercial spent nuclear fuel and government-owned high-level radioactive wastes are intended to be disposed of in the repository at Yucca Mountain, NV. Transuranic wastes are intended to be disposed of in the Waste Isolation Pilot Plant near Carlsbad, NM, which has been constructed and is just now operating. Low-level radioactive wastes owned by the Department of Energy (DOE) are disposed of at various DOE sites. Commercial low-level wastes are to be disposed of by the states in regional disposal sites, only two of which exist at the present time (at Hanford, WA, and Barnwell, SC). Attempts at siting other regional commercial low-level waste burial grounds have resulted in failure (for example the Midwest Compact Site intended to serve Ohio, Indiana, Iowa, Minnesota, Missouri, and Wisconsin) or have been impeded by controversy and legal action (for example the Ward Valley, CA, site intended to serve Arizona, California, North Dakota, and South Dakota).

CHAPTER 3 REFERENCES

Abelson, Philip H. "Chlorinated Dioxins," *Science*, 220, June 24, 1983.

Abelson, Philip H. "Excessive Fear of PCBs," *Science* 253, 361, June 24, 1991.

Charlson, R. J. et al., *Science* 255, 423, 1992.

Cooney, Catherine M. *Environmental Science and Technology* 31, 269A, June 1997.

Environmental Protection Agency (EPA). "Draft EPA Mercury Study: Report to Congress," EPA-452/R-96-001a, June 1996.

Environmental Protection Agency (EPA). *Federal Register*, 62, 24887, 1997.

Environmental Science and Technology (ES&T) 12, 507A, December 1994.

Hamill, Patrick, and Toon, Owen Brian. *Physics Today*, p 34, December 1991.

Karl, Thomas R., Nicholls, Neville, and Gregory, Jonathan. "The Coming Climate," *Scientific American*, p 78, May 1997.

Ladbury, Ray. "Model Sheds Light on a Tragedy and a New Type of Eruption," *Physics Today*, p 20, May 1996.

Lide, D.R. editor, *CRC Handbook of Chemistry and Physics*, 75th edition, CRC Press, Inc., 1994.

Likens, G.E., Driscoll, C.T., and Buso, D.C. *Science*, 272, 244, April 12, 1996.

Molina, Mario J., and Rowland, F. Sherwood. *Nature*, 249, 810, 1974.

Office of Technology Assessment (OTA). *Beneath the Bottom Line: Agricultural Approaches to Reduce Agrichemical Contamination of Groundwater*, OTA Brief, November 1990.

Pacific Northwest National Laboratory (PNNL). *Hanford Site 1996 Environmental Report*, PNNL-11472, 1997.

Tschirley, Fred H. "Dioxin," *Scientific American* 254, 29, February 1986.

Zakrzewski, Sigmund F. *Principles of Environmental Toxicology*, American Chemical Society, 1991.

Zurer, Pamela. *Chemical and Engineering News*, p 7, January 7, 1991.

CHAPTER 4

Pollution Reduction Technologies

Environmental remediation technologies include technologies for reducing pollution from existing sources (mainly gaseous and liquid waste streams in the past, but now also solid waste streams) and technologies for removing pollutants from inactive hazardous waste sites and contaminated ground water. Pollution prevention, of course, eliminates the need for pollution reduction and pollutant removal.

The difference between a pollution reduction technology and a pollutant removal (cleanup) technology is often merely a matter of classification: one person's pollution reduction technology is another person's cleanup technology and vice versa. Nevertheless, a pollution reduction technology can be defined as one that is directed at an air emission source, a water effluent source, or a solid waste stream and serves to capture an unwanted pollutant before it reaches the environment. A cleanup technology then becomes a technology that is applied to soil or water after the water or soil has become contaminated. Cleanup technologies are not, in general, applied to air masses, but pollution reduction technologies may be applied to an air stream that is part of a soil or water cleanup technology. Pollution control or cleanup technologies are sometimes grouped as physical, chemical, thermal, biological, or solidification/stabilization technologies (EPA 1990).

4.1 AIR POLLUTION REDUCTION TECHNOLOGIES

Moore's second law of pollution control is, "Don't let it get out in the first place." And one of the corollaries of that law is, "Provide controls so that the pollutant does not escape." Current control technologies are discussed in this section.

Air pollutants include particulate matter, volatile organic compounds (VOCs), acid gases, and miscellaneous chemicals including radionuclides. Particulate matter is removed by filters, scrubbers, centrifugal separators, or electrostatic precipitators; VOCs are removed by incineration, condensation, or adsorption on activated charcoal; acid gases are removed by alkaline scrubbers; and other chemicals are removed by adsorption or by scrubbing in appropriate chemical solutions. Heavy metals often occur as components of flyash and can be removed by particulate removal technologies. The specific technology to be used depends on the nature of the pollutant, flow rates, concentrations, and temperature. Often a blower is required to move (push or pull) the air mass through the device.

4.1.1 Particulates (Moore et al. 1984)

Particulates can be removed from an air stream by several methods. Special handling of the captured waste is often required if the waste includes hazardous or radioactive substances.

High Efficiency Particulate Air (HEPA) Filters. HEPA filters (Figure 6) are usually constructed of matted fibers held in a frame, like a furnace filter or automobile air filter. The frames can be of any dimension suitable for installation in series or in parallel in the air stream of interest. They can be 99.97 percent efficient for particles greater than 0.3 micron in diameter. HEPA filters are frequently used to remove radionuclides from an air stream when the radionuclides are present as particulates.

Baghouses. Baghouses (Figure 7) are made of cloth (like vacuum cleaner bags) and can be 99.8 percent efficient for particles above 0.5 micron in diameter. Baghouses are also used to remove radionuclides when they are in particulate form.

Sand Bed Filters. Sand bed filters (Figure 8) can measure 100 ft × 100 ft × 7 ft deep and can be 99.97 percent efficient for removal of

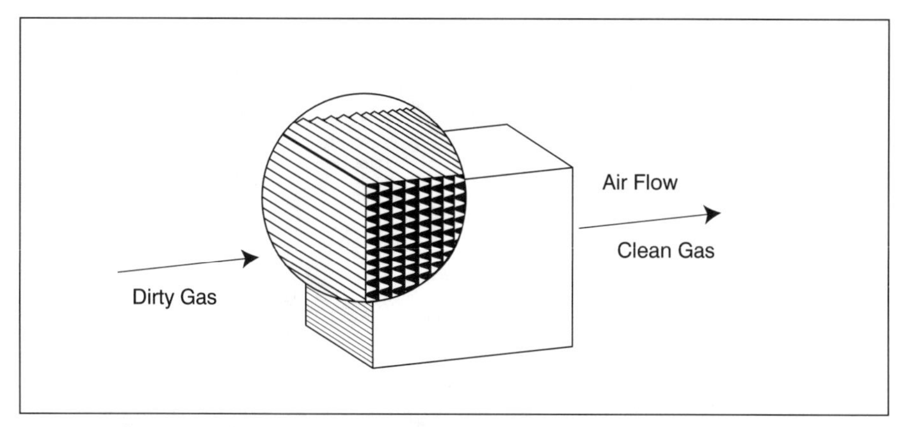

Figure 6. *High efficiency particulate air filter (Moore et al. 1984).*

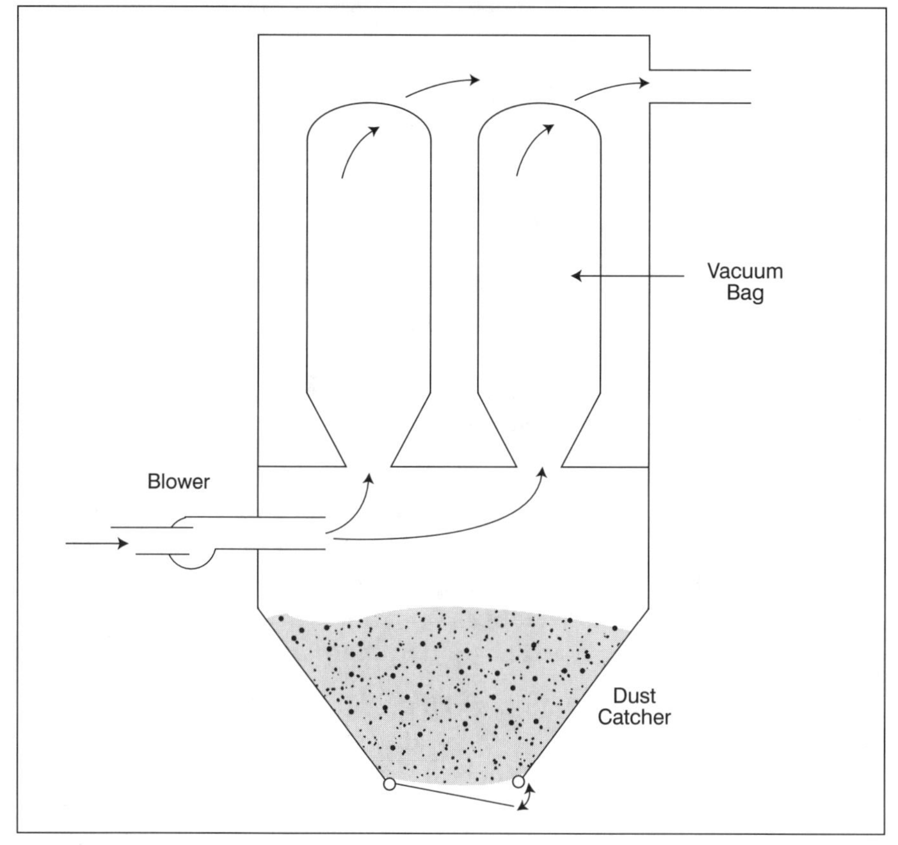

Figure 7. *Baghouse filter.*

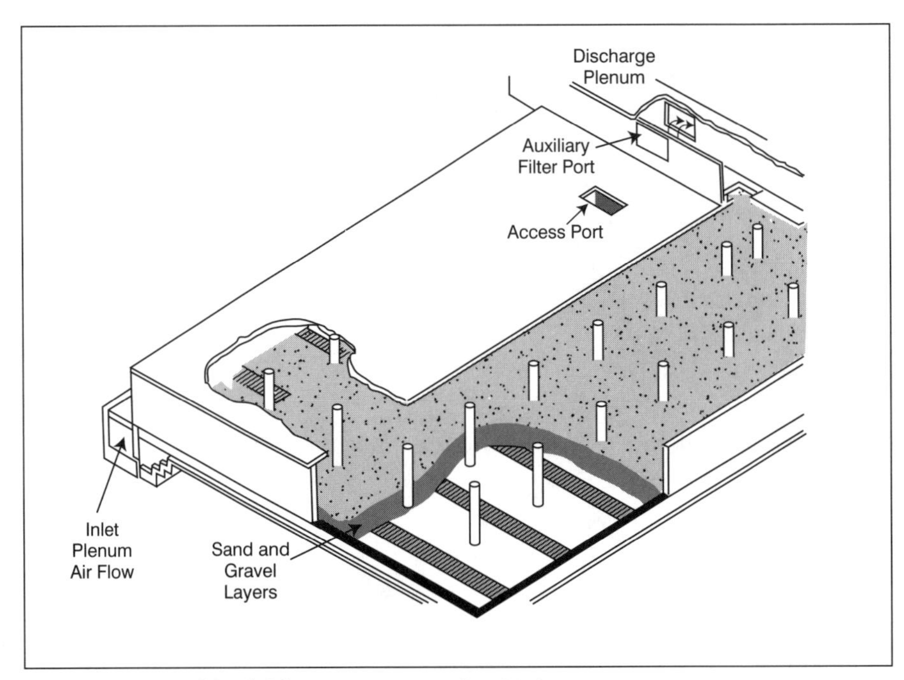

Figure 8. *Sand bed filter (Moore et al. 1984).*

particulates from an air stream. Sand bed filters are also used to filter liquids (usually water). For an air stream, the particulate-laden air ascends through the filter, and for a water stream, the particulate-laden water descends through the filter.

Electrostatic Precipitators. Electrostatic precipitators (Figure 9) can be up to 99 percent efficient for charged particles above 0.3 micron in diameter. The precipitator provides a high voltage (60 kV, for example) that attracts the charged particles to the plus or minus electrode. Sometimes a chamber is provided in advance of the high voltage electrodes to place a charge on the particles.

Wet Scrubbers. Wet scrubbers (Figure 10) can be static, inertial, countercurrent, or spray. In the static scrubber, the air is forced through a vat of water. In the inertial scrubber, the air stream and the water travel together in a centrifugal path. The countercurrent scrubber is a packed column in which the water passes downward and the air passes upward. In the spray scrubber, a water spray is introduced, and the air and water travel in opposite directions.

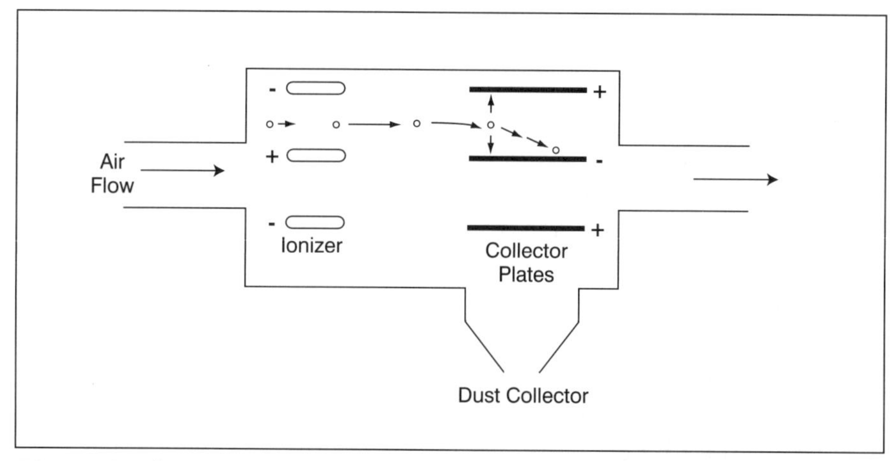

Figure 9. *Electrostatic precipitator (Moore et al. 1984).*

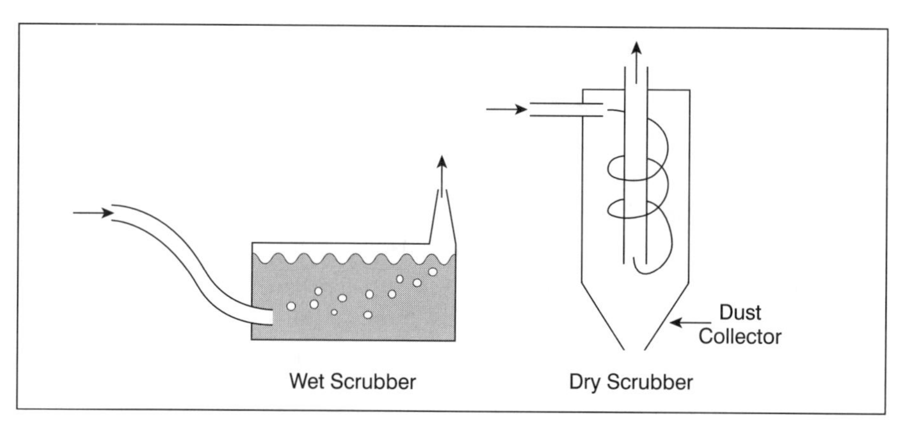

Figure 10. *Wet and dry scrubbers.*

Dry Inertial Devices. Dry inertial devices (Figure 10) are usually cyclone (centrifugal) separators that force the particles to the outside of the chamber. Inertial devices can be effective with particles down to 10 micron in diameter.

4.1.2 **Volatile Organic Compounds**

VOCs can be removed from an air stream by incineration (simple incineration at 1600° to 2000°F or catalytic incineration at 1000° to

1200°F with a Pt or Pd catalyst), by condensation, or by adsorption on activated charcoal.

Thermal Treatment of Gases (Incineration). Thermal treatment of an offgas is for the purpose of oxidizing (burning) an organic pollutant. The pollutant may be from an industrial process or may have been stripped by a carrier gas (usually air) from a liquid or a solid (often soil). Simple flaring of the gas (burning at the end of a pipe) or burning in a chamber can be used. The major end products are intended to be carbon dioxide and water. If, however, there are constituents in the pollutant other than carbon, hydrogen, oxygen, and nitrogen (such as chlorine), then care must be taken to avoid undesirable products (such as HCl).

Condensation. Passing an air stream through a condenser cooled by, say, liquid nitrogen will condense many organic vapors (gases) to either the liquid or solid phase (Figure 11).

Adsorption on Activated Charcoal. Activated charcoal is finely divided charcoal (carbon) that has a large surface area and is initially without impurities adsorbed on the surfaces. Activated charcoal is an excellent adsorber for organic compounds both from an air stream and from water. "Spent" charcoal can be reactivated by heating or by passing steam over the charcoal.

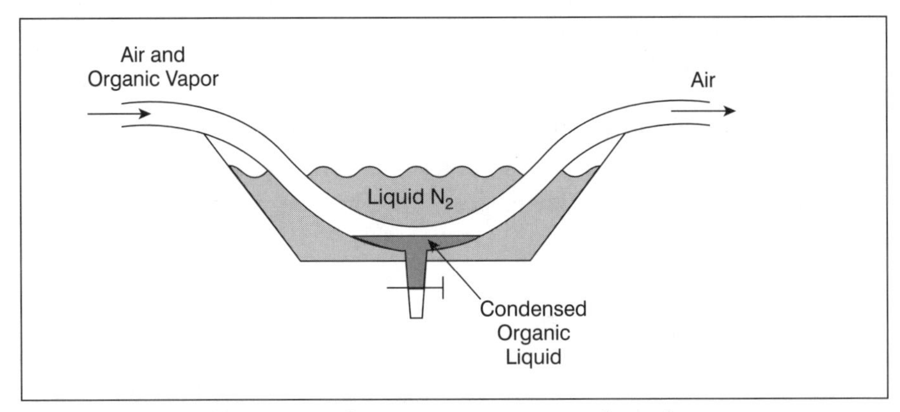

Figure 11. *Condensation of an organic vapor to liquid in an air stream.*

4.1.3 **Chemicals, Including Radionuclides**

The separation of a chemical from an air stream is based on chemical properties. The separation of a radionuclide from an air stream is also based on the radionuclide's chemical properties, and not on its nuclear properties.

Acids. Neutralization with a base (NaOH for example) in a scrubber is the usual method of removing an acid. (See HCl below.)

I-129 and I-131 (fission products). Isotopes of iodine may be precipitated with silver nitrate in a reactor or precipitator. The process is 99 percent efficient. A charcoal filter can also be used to remove iodine with 99.9 percent efficiency (Moore et al. 1984).

Sulfur Dioxide. SO_2 comes mostly from coal-fired power plants, although it often arises as H_2S from industrial processes and from decaying organic matter, which is then oxidized to SO_2 and to SO_3 in air. Sulfate aerosols can be formed. High-sulfur coals can be washed before burning to remove some sulfur. Scrubbers that precipitate $CaSO_4$ from a solution of $CaCO_3$ (limestone) are frequently used. Sludge disposal of the $CaSO_4$ then becomes a problem.

Oxides of Nitrogen. Oxides of nitrogen include N_2O, NO, and NO_2. N_2O originates from natural sources. The other oxides of nitrogen originate from high temperature combustion processes (reaction of nitrogen and oxygen from the air). Amounts may be reduced in an air stream by reducing the temperature of combustion, for example by using a fluidized-bed combustor. A fluidized-bed combustor that injects limestone can reduce emissions of both SO_2 and NO_2. Reduction of SO_2 occurs from the reaction of the SO_2 with $CaCO_3$, and reduction of NO_2 occurs from the lower temperature of combustion. The lower temperature of combustion, however, also reduces the thermodynamic efficiency of the conversion of heat to electricity, if the furnace is used for that purpose. Catalytic converters can be designed to lessen the amount of NO_x emitted by reducing NO_x to N_2 and O_2 and can also be designed to oxidize hydrocarbons and CO to CO_2 and H_2O. A rhodium catalyst will oxidize CO to CO_2 and reduce NO to N_2 in the same reaction: $CO + NO = CO_2 + 1/2N_2$ (Friend 1993). Another catalytic reduction of NO occurs with ammonia and oxygen: $NO + NH_3 + 0.25O_2 = N_2 + 1.5H_2O$ (Durrani 1994).

This reaction is sometimes used to remove oxides of nitrogen from an air stream.

CFCs and Halons (fluorocarbons that contain bromine). The main way of eliminating CFCs and halons from the atmosphere is to stop their production and use. Complete phaseout of production of CFCs, halons, and carbon tetrachloride by 2000 is required under the CAAA. CFCs were to have been phased out by 1995 by executive order. CFCs used as refrigerants are being replaced with HFCs and HCFCs (H = hydrogen), which will be phased out later. Thermal destruction of CFCs must be carried out carefully because of the production of phosgene. A chemical method for destruction of CFCs involves reduction of the CFC to NaCl and NaF in a bath of sodium metal dissolved in anhydrous liquid ammonia.

HCl. HCl results from some industrial processes and from the incineration of chlorinated organic compounds. HCl is captured in NaOH scrubbers, which can be as simple as a water solution of NaOH through which the air stream is bubbled (Figure 10), or which can be more complicated counter-current packed-bed scrubbers (Figure 12). Removing HCl is important after the incineration of chlorinated organic compounds.

CO. Automobile catalytic converters are intended to catalyze the oxidation of CO to CO_2.

Lead. Lead in the atmosphere is being reduced by removing leaded gasoline from the market. (Lead-based paints should not be used, particularly where children might chew on a painted object.)

Chromium, Arsenic, Zinc, Silver, etc. Heavy metals,

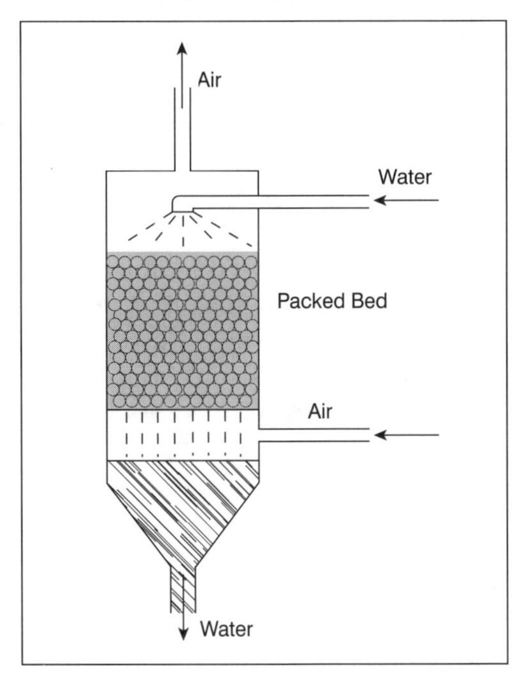

Figure 12. *Counter-current, packed-bed scrubber.*

such as chromium, arsenic, zinc, and silver (usually in the form of metal oxides), in an air stream as particulates, can be removed by dry filtration techniques. Trace metals from coal-fired power plants are often in the fly ash, but can also be in the bottom ash, which then requires some kind of containment to prevent leaching from the ash pile. If not in particulate form, special chemistry is needed.

4.2 WATER POLLUTION REDUCTION TECHNOLOGIES

Water pollution reduction technologies apply to the treatment of domestic sewage, industrial discharges, and agricultural runoff.

4.2.1 Sewage Treatment

Primary sewage treatment (Figure 13) means removal of solids (particles) by gross filtration, sedimentation, and/or coagulation with alum ($KAl(SO_4)_2$). Secondary sewage treatment means destruction of biological contaminants and organic compounds (removal of biological oxygen demand) by aeration of the water and bacterial action in activated sludge. Tertiary sewage treatment means removal of other constituents such as phosphates, nitrates, and heavy metals by special techniques. Both aeration and bacterial action oxidize organic compounds. Heavy metals that are not removed from sludge may make the sludge unsuitable for land disposal. Good secondary treatment can remove 90 percent of biological oxygen demand (BOD) from the water. Activated charcoal adsorption in a tertiary treatment can remove much of the rest of the BOD, i.e., organic compounds. Other tertiary treatments include removal of toxic chemicals, removal of phosphates with Al or Fe salts to form a phosphate precipitate, and removal of nitrates by bacterial reduction to N_2. Any primary, secondary, or tertiary treatment of sewage is almost always followed by the addition of chlorine for disinfection purposes.

4.2.2 Industrial Discharges

Industrial discharges in the past contained all manner of pollutants, frequently heavy metals, but also organic compounds. Now, industrial discharges are regulated by the Clean Water Act and must meet the requirements of the discharger's National Pollutant Discharge

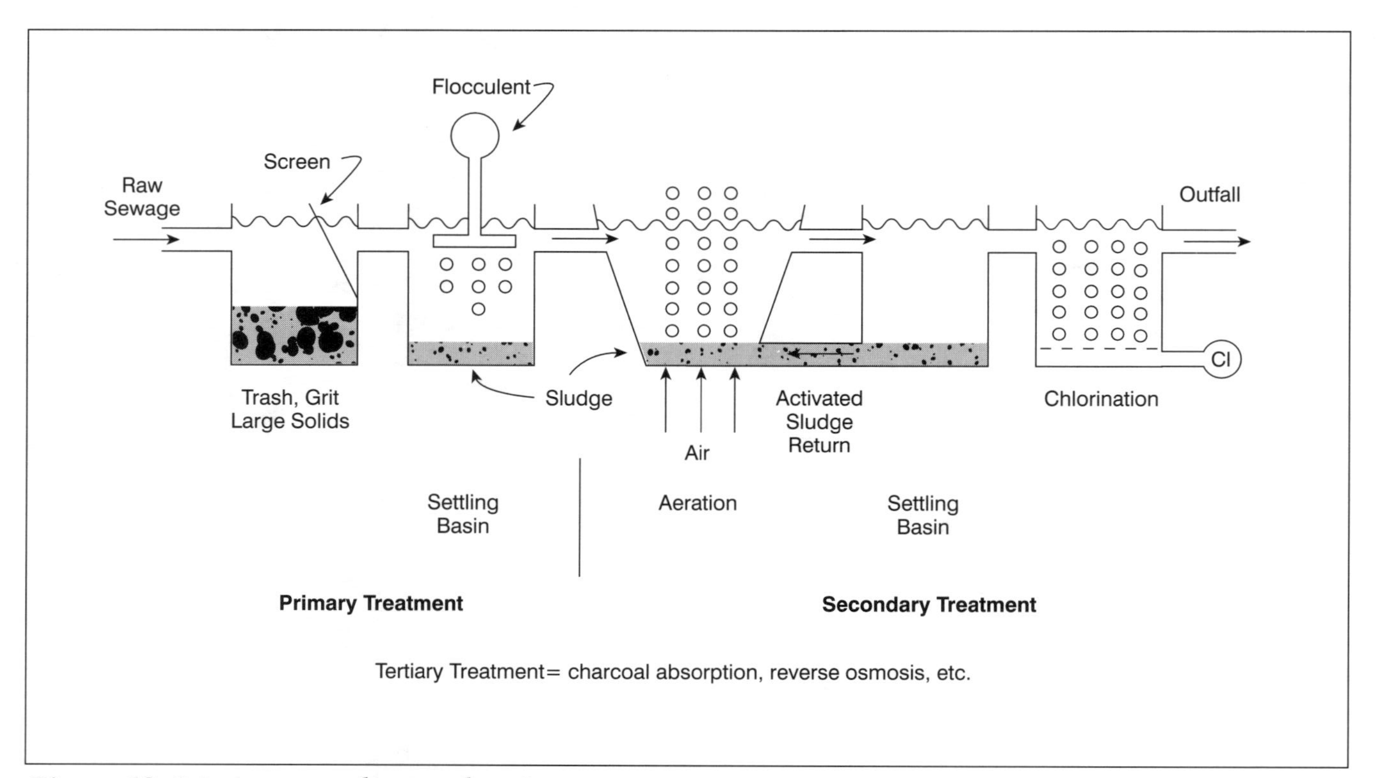

Figure 13. *Primary, secondary, and tertiary sewage treatment.*

Elimination System (NPDES) permit. Organic compounds are captured or destroyed by incineration, heavy metals are removed by precipitation or ion exchange, viruses and bacteria are destroyed, etc.

4.2.3 Agricultural Practices

Reduction of pollutants in agricultural runoff is best effected by reducing the use of nitrate fertilizers and by reducing the amounts, mobility, human toxicity, and environmental persistence of pesticides.

4.2.4 Storm Water Treatment

Storm water problems include raw sewage overflow from sanitary sewers into which storm water has been introduced. Some storm waters are now regulated by NPDES permits. These discharges include point discharges of storm water into waters of the United States from industrial, city, and county facilities and from construction projects over five acres in size. An NPDES permit could require treatment of storm water, although such treatment is not yet widespread.

4.3 SOLID WASTE REDUCTION TECHNOLOGIES

Solid waste reduction technologies can reduce both the volume (compaction) and the mass (incineration) of solid waste, thereby reducing the amount of eventual landfill space required for final disposal.

4.3.1 Separation and Recycling

While some automated mechanical separation facilities exist for solid wastes, most of the separation must still be done by hand by the generator. Once the separation is effected, say, into paper, metals, glass, and different kinds of plastics (milk bottles, pop bottles), then the materials can be recycled. Ferromagnetic and paramagnetic materials can be separated by magnets. A problem with recycling is the development of adequate markets for the recycled materials.

4.3.2 **Compaction**

Compaction will reduce the volume of solid waste but not the mass. Also, no separation into components is effected.

4.3.3 **Incineration**

Incineration will reduce both the mass and volume of solid waste. Municipal incinerators were not uncommon before 1970 when air pollution control devices were not required. But the requirements of the Clean Air Act (1970) made landfills more attractive than incinerators. Now, with increasingly strict solid waste disposal regulations (RCRA 1976 and the land disposal restrictions in HSWA 1984), incinerators are once again of interest to municipalities. But the new incinerators may require filters to capture particulates and scrubbers to capture acid gases. One advantage of municipal incinerators is the opportunity to capture steam for industrial or space heating and for generating electricity. Municipal waste incinerators must be distinguished from hazardous waste incinerators. The latter have more stringent regulatory requirements. In early 1993, only 18 commercial hazardous waste incinerators were operating in the U.S. (See Sec. 5.3.1.)

4.3.4 **Disposal in Landfills**

Disposal of municipal garbage in landfills is still allowed, but the land disposal of hazardous waste (under RCRA) is now highly regulated (see Chapter 6).

4.3.5 **Composting**

Composting of yard and agricultural wastes is a method of recycling and/or reducing both the mass and volume of these wastes (to zero if they are reapplied to the soil).

CHAPTER 4 REFERENCES

Durrani, Sher M. "The SNOX Process: A Success Story," *Environmental Science and Technology* 28, 88A, 1994.

Environmental Protection Agency (EPA), U.S. *The Superfund Innovative Technology Evaluation Program, Progress and Accomplishments Fiscal Year 1989*, EPA/540/5-90/001, March 1990.

Friend, Cynthia M. "Catalysis on Surfaces," *Scientific American*, p 73, April 1993.

Moore, E.B. et al. *Control Technologies for Radioactive Emissions to the Atmosphere at U.S. Department of Energy Facilities*, Pacific Northwest Laboratory, PNL-4621, 1984

CHAPTER **5**

Pollution Cleanup Technologies

Cleanup technologies are applied to contaminated water and contaminated soils. Cleanup (pollution control) technologies are also applied to air streams that result from the processes of cleaning contaminated water or soil. The useful technologies, as noted in Chapter 4, may be classified as thermal, biological, chemical, physical, or solidification and stabilization technologies (EPA 1990, Kopstein 1989). Clearly there is overlap between and among these classifications, and the boundaries are not hard and fast. For example, heating (thermal) a liquid waste stream may result in distillation (physical) and/or oxidation (chemical) of the pollutant. Nevertheless, these classifications can be useful for understanding the various technologies. Cleanup technologies may also be classified as in situ or remove-and-treat technologies. The science of cleanup technologies is, for the most part, well known; but the application of the science is still evolving.

5.1 AIR STREAM CLEANUP TECHNOLOGIES

Air stream cleanup technologies are discussed in Sec. 4.1 (i.e., particulate removal, incineration, condensation, adsorption, and chemical reactions). Small amounts of air can be cleaned to any desired level of purity. But all of the atmosphere, obviously, cannot be cleaned. The best air cleanup technology is not to allow the pollutant to escape in the first place (Moore's second law of pollution control).

5.2 WATER CLEANUP TECHNOLOGIES

The term "water cleanup" is usually understood to mean cleaning up ground water that has been polluted in some manner, perhaps by contact with hazardous waste, by underground injection of waste (such as CCl_4), or by seepage of polluted liquids into the aquifer. Water may be pumped to the surface, treated, and reinjected or discharged elsewhere. Water may also be treated in situ, depending on the characteristics of the pollutant. The technologies discussed in Section 4.2 are applicable here. Water pumped to the surface may be treated to any desired degree of purity, i.e., for drinking water purposes; but the aquifer itself may, in many cases, never be cleaned to an adequate degree of purity, either by pump-and-treat or by in situ technologies.

5.2.1 Pump (physical) and Treat (chemical or physical)

Pump-and-treat (Figure 14) is the standard technology for removing pollutants or contaminants from aquifers. In this technique, ground water is pumped to the surface, the water is treated at the surface, and then is pumped back into the ground or discharged elsewhere. Bulk amounts of contaminants can be removed from the aquifer, but the technology may not be effective in cases where the contaminants are strongly held by soil particles. Chemical species may be adsorbed on soil particles and liquids may be trapped in small interstices by capillary forces (surface tension). Very long pumping times may be required to remove the contaminants adsorbed on the soil particles and bring them to the surface in the water for treatment. Surfactants may be needed to remove the liquids held by capillary forces. Also, the pump-and-treat technology is not effective in cases where slightly soluble or insoluble liquid contaminants lie in pockets on top of or below the aquifer, depending on whether or not the liquid contaminant is lighter or heavier than water. The pumped stream, of course, can be cleaned to any desired degree of purity, say, for drinking water purposes. It is the aquifer that cannot be completely cleaned. Some observers now flatly state that pump-and-treat technologies do not

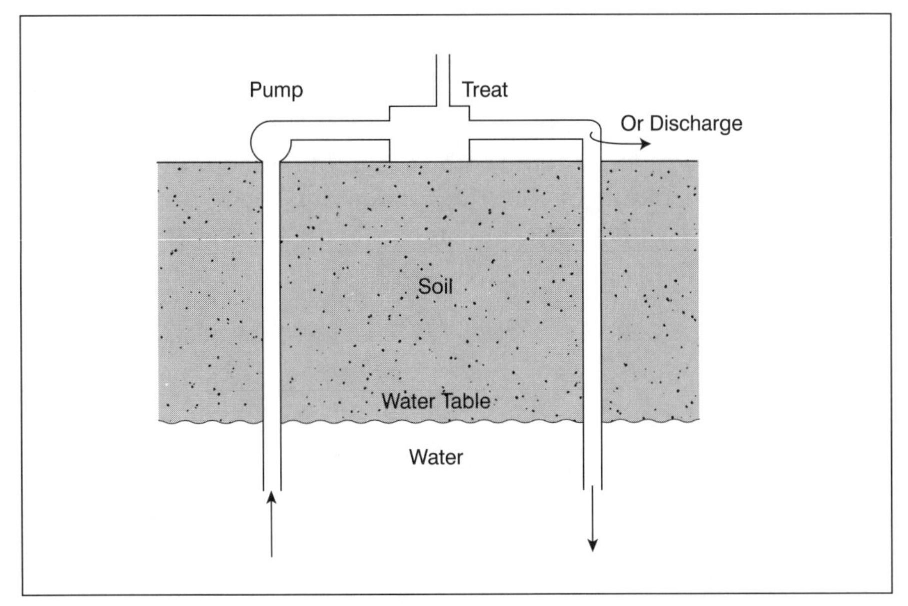

Figure 14. *Pump-and-treat technology.*

work. As stated above, this is because not all of the contaminant can be brought to the surface within a reasonable time. Pumping can go on for a very long time. Or pumping can be stopped only to have the operators find later that the concentration of contaminant in the well water has risen again, because of desorption of the contaminant from soil particles upgradient of the well, or because of additional dissolving in the water of nearly insoluble liquids. The "treat" part of pump and treat includes filtration, flocculation (coagulation), precipitation, as well as the technologies discussed below.

5.2.2 **Air Stripping of Volatile Organic Compounds from Water (physical)**

VOCs dissolved in or carried by water can be partially removed from the liquid phase by passing air through the liquid. Usually, the warmer the air, the better. Both water vapor and some of the VOC will be removed by air stripping. Subsequent processing of the air stream is required to remove the organic constituent and to separate it from the water vapor. Subsequent processing could include adsorption of the VOC on activated charcoal, condensation, or incineration. Water may be treated in situ by this method. In one example, two wells are sunk, one into the aquifer and one above the aquifer. Air is forced into the well in the aquifer and withdrawn through the well above the aquifer, usually with the aid of a vacuum (Figure 15). Subsequent separation of the VOC and water vapor takes place at ground level. As with pump and treat, in situ treatment may never complete the job.

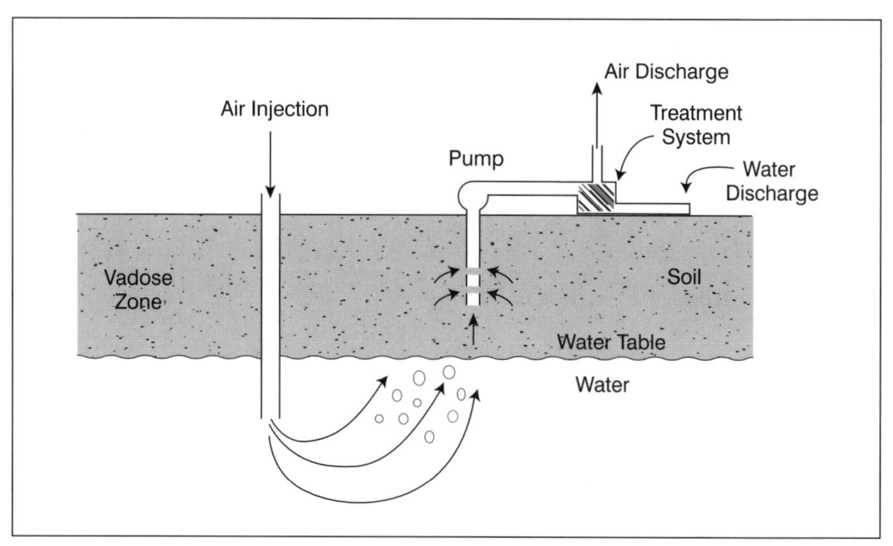

***Figure 15.** Air stripping of a volatile organic compound from water in an aquifer.*

5.2.3 **Thermal Treatment of Liquids (thermal)**

Thermal treatment of liquids usually means distillation (Figure 16) of water that contains VOCs. If the unwanted constituent is not volatile (such as an inorganic salt containing a metal ion), then water will be distilled away, leaving behind a solid containing the non-volatile constituents of the solution. If, on the other hand, the unwanted constituent is a soluble and volatile organic compound, then distillation of both water and the organic compound will take place; and the unwanted organic compound will partition itself between the liquid and vapor phases. If the organic compound has a higher vapor

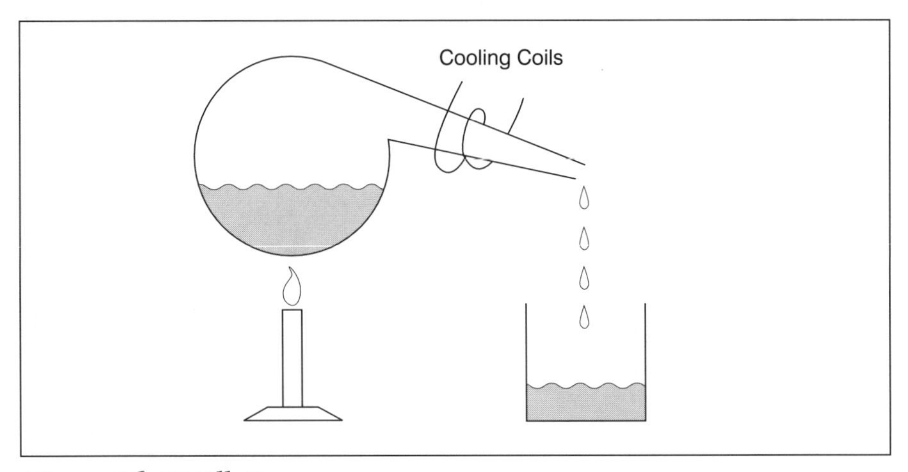

Figure 16. *Distillation.*

pressure than water (i.e., lower boiling point), then the organic compound will become more concentrated in the vapor phase than in the liquid phase. In this way, a separation of water and the organic compound is effected. Multiple distillation stages will be required to effect a useful separation.

5.2.4 **Filtration, Adsorption (physical)**

Sand bed filters and charcoal adsorption are applicable to the cleanup of ground water. These technologies may be used in the treatment of potable water (Section 5.2.12) and may be used in the pump-and-treat process.

5.2.5 **Membrane Technologies (physical)**

Membrane technologies can be effective in removing almost all pollutants. Membranes are available with different pore sizes that will pass water only, or water and small molecules, but not larger molecules. In osmosis (Figure 17), a semipermeable membrane (a membrane with the smallest pore size) is used, which means that only water molecules penetrate the membrane. In ordinary osmosis, pure water diffuses through the membrane to dilute the solution on the other side. In reverse osmosis, a pressure is applied on the solution,

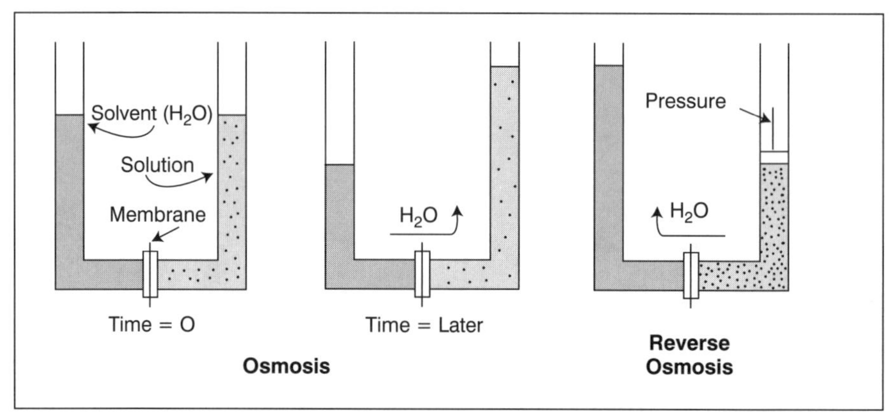

Figure 17. *Osmosis, reverse osmosis.*

which forces pure water back through the membrane to the pure water side, leaving behind an increasingly concentrated solution. (The solution contains dissolved solute.) Membranes with larger pore sizes are useful for ultrafiltration or microfiltration. Depending on the pore size, dissolved solutes, viruses, bacteria, pollens, and/or particles will not be able to pass through the membrane. Thus, small molecules may pass through a membrane while leaving larger ones behind. Dialysis and electrodialysis are flow techniques that employ permeable membranes, which will pass water as well as inorganic ions and small organic molecules, but not large organic molecules such as proteins. Various combinations of membranes and electric fields can selectively remove the ions and, therefore, purify the liquid. In electrodialysis, both small ions (charged atoms or molecules) and water pass through the membrane, leaving proteins behind, but the ions move faster through the membrane than water because of the attractive force of the electric field.

5.2.6 **Solvent Extraction (physical)**

Solvent extraction occurs when an organic compound is soluble both in water and in an organic solvent that is immiscible with water. The organic compound partitions itself between the two solvents and is thus removed from the water. Repeated stages are necessary.

5.2.7 **Ion Exchange (chemical)**

Ion exchange compounds such as zeolites can be used to trap heavy metal ions, thus removing the metal ions from solution, while releasing sodium or potassium ions to the solution. As in the case of home water-softener units, the ion exchange compound must be periodically recharged to remove the heavy metals and replace them with sodium or potassium ions.

5.2.8 **Freeze Crystallization (physical)**

Partially freezing a dilute solution containing dissolved organic or inorganic constituents will separate pure water (ice crystals) from the rest of the solution containing the solute. Thus, the solution becomes more concentrated in impurities.

5.2.9 **Photolysis, Oxidation, and Catalysis (physical and chemical)**

Various combinations of photolysis (using sunlight as well as other visible and ultraviolet light sources), of oxidizing agents (such as hydrogen peroxide), and of catalysts (such as titanium dioxide) have been used to convert organic compounds, including gasoline, into carbon dioxide, water, and HCl. As a replacement for chlorination (in order not to form chlorinated organic compounds in treated water), photolysis can be used in the range of 240 to 280 nm for water disinfection purposes. There are, however, concerns about the effectiveness of UV treatment. Also, note that incomplete photocatalytic oxidation of TCE or PCE, i.e., with UV and TiO_2, can produce phosgene and Cl_2.

5.2.10 **Electron Beam Irradiation (chemical)**

Electron beam irradiation has been used to treat potable water and water effluent from secondary sewage treatment. The technique is effective in removing (destroying) some organic compounds such as TCE, carbon tetrachloride, tetrachloroethylene, and benzene. It is now being tried on PCBs.

5.2.11 **Supercritical Oxidation**

High temperature and high pressure (500°C and 300 atm) can make some organic compounds more soluble in water and more easily oxidized.

5.2.12 **Potable Water Treatment**

Potable water treatment includes filtration (sand filters), coagulation with alum, adsorption by activated charcoal, disinfection with chlorine and/or UV and, occasionally, distillation. Inorganic compounds can also be removed by ion exchange, electrodialysis, and/or reverse osmosis. For example, removal of inorganic ions is important in the preparation of potable water from sea water.

5.3 **SOIL CLEANUP AND HAZARDOUS WASTE TREATMENT TECHNOLOGIES**

Soil cleanup technologies include both remove-and-treat technologies and in situ technologies. Soils can be cleaned by both technologies to reasonable levels, except that it is often difficult to remove metals and some radionuclides. Sediment cleanup is sometimes referred to as "suck, muck, and truck." Hazardous waste treatment technologies are often the same as soil cleanup technologies, so both are discussed here.

5.3.1 **Thermal Treatment of Solids, i.e., Soils or Hazardous Wastes (thermal)**

Thermal treatment of solids (soils or hazardous wastes) is heating a solid to vaporize (drive off) a volatile organic compound, decomposing or oxidizing (directly decomposing or reacting with oxygen, i.e., incinerating) complex organic compounds to form simple organic compounds (ideally carbon dioxide and water), or solidifying (vitrifying) the solid to trap a nonvolatile pollutant such as a heavy metal. Particulates or compounds such as HCl may be undesirable byproducts of thermal treatment. Low temperatures (150°C to 400°C) may remove a volatile organic compound from soil without decomposing it. An externally fired rotary kiln can be effective in removing volatile organic compounds from soil. Higher temperatures (over 900°C) will decompose or oxidize organic compounds. Temperatures above 1100°C will vitrify soils. Thermal treatments ordinarily do not remove metal ions from soils, although vitrified soils will immobilize metal ions. Depending on the application, the source of heat can be a flame, infrared radiation (up to 1000°C), a plasma arc (usually for gases), or electric resistance heating. Vitrification and thermal desorption are in situ or remove-and-treat technologies for soils, but the other thermal technologies require removing, treating, and replacing the soil.

Thermal Desorption. Heating soils at low temperatures can drive off VOCs. The heating can be done both in situ and in low-temperature incinerators. Offgas treatment is necessary to remove the VOC from the offgas air stream.

Incinerators. The most common thermal treatment device is the incinerator, which in its most simple form is just a furnace. Additional fuel may be added, and various scrubbers and filters may be needed to capture certain offgases and particulates. Alkaline scrubbers may be required to remove acid gases. Baghouses, HEPA filters, and/or electrostatic precipitators may be required to remove particulates. A fluidized bed construction may improve some incinerator applications. A fluidized bed incinerator is one in which a stream of air is forced upward through the grate to lift the burning fuel and waste and sometimes a bed of sand above the grate. This results in better mixing of waste, fuel, and oxidant. A rotary kiln incinerator is an incinerator that has a horizontal, but slightly downward tipped, cylindrical burning chamber. The chamber rotates on its axis, thereby tumbling the waste (and any solid fuel) to be burned. If the rotary kiln only desorbs organic compounds, then another device such as a second combustion chamber is required to remove the organic compounds from the air stream. If chlorinated organic compounds are burned, care must be taken to avoid release of the combustion product HCl. Dioxins are also created in some combustion processes when chlorinated organic compounds are burned. Plasma heating or heating by infrared radiation can be used, as well as heating by burning fuel. In a plasma system, the waste contacts the ionized plasma gas. In an incinerator, organic compounds are converted to carbon dioxide and water and metals are converted to metal oxides. Sometimes incomplete combustion of organic material is indicated by CO as a product. In an incinerator in Spokane, WA, SO_2 is a combustion product from wallboard ($CaSO_4$) and tires. Volume reductions in an incinerator of 100 to 1 are possible. Heat recovery is also a possibility.

Soil incineration in a rotary kiln was the method of choice for removing dioxins from large quantities of soil at Times Beach, MO. Dioxin was vaporized in the kiln and then oxidized in a secondary combustion chamber (Hileman 1997). An incinerator to burn mixed (radioactive and hazardous) wastes is located at Oak Ridge, TN. These wastes are DOE wastes containing uranium. Both RCRA and TSCA (for PCBs) permits were obtained. A 99.9999 percent burn efficiency is claimed. RCRA prohibits indefinite storage of this mixed waste without treatment, i.e., incineration, before disposal. After a long and controversial planning, construction, and testing process that included lawsuits and even became an issue in the 1992 presidential election, a hazardous waste incinerator at East Liverpool, OH, was approved for full commercial operation by EPA in 1997 (ES&T 1997).

Incinerator slag may not pass the toxicity characteristic leaching procedure (TCLP), in which case the slag is still a hazardous waste.

Recently EPA has backed away from incineration as the treatment of choice for some land-disposal restrictions (LDRs). In 1993, EPA proposed alternatives to combustion as a treatment for contaminated soils (58 FR 48092, September 14, 1993).

Vitrification. Vitrification consists of mixing glass with the waste to be vitrified and melting the mixture under a high temperature to form a glassified mass. For above-ground applications, the molten glass is poured into a suitable container, often stainless steel, to solidify. Vitrification is being used at Savannah River to contain DOE's high-level radioactive wastes.

In Situ Vitrification. In situ vitrification is carried out by inserting electrodes into the soil, assuring that the soil will conduct the initial current before melting begins, and passing a large electric current through the electrodes to melt and glassify the soil. While heavy metals may be vitrified along with the soil, organic compounds may be vaporized and must be trapped. Obviously, this technology works best when the soil contains large amounts of sand (SiO_2).

5.3.2 **Bioremediation (biological)**

Bioremediation (Atlas 1995) means using microorganisms to treat or degrade hazardous wastes, i.e., hazardous organic substances. These substances may be liquids or damp solids. The microorganisms include bacteria, algae, and fungi. A compost pile is a common, everyday example. Bioremediation may be done in situ or in a bioreactor that handles a liquid, slurry, or solid. The bacteria may be aerobic or anaerobic. Some 400 different microorganisms have been identified that may be effective in bioremediation. The microorganisms synthesize enzymes that break down the complex organic compounds to water, carbon dioxide, chloride ion, and methane. The pH, temperature, oxygen supply, and nutrients must be controlled. Bioremediation has been used on petroleum products, and more recently on TCE. Care must be taken, however, because PCE and TCE have been observed to decompose to vinyl chloride, which is a carcinogen. PCBs can be biodegraded by both aerobic and anaerobic bacteria. Parts per million is the concentration range in which bioremediation is effective. The end products can be carbon dioxide, water, and chloride ion.

At first it was reported that naturally occurring bacteria, when applied judiciously with nitrogen nutrients, were effective in removing oil from the shorelines of Prince William Sound, AK, following the Exxon Valdez oil spill. Maybe. Later indications are that perhaps the physical force from the high pressure wands used to apply the solution did much of the cleaning (Stone 1992).

5.3.3 **Phytoremediation (biological)**

While many plants are damaged and killed by high concentrations of metal ions in soils, some plants are known that will thrive under these conditions and will actually absorb large quantities of metal ions from the soil. The plants can then be harvested and processed for their metal content. Alpine pennycress (*Thlaspi caerulescens*) will absorb zinc, cadmium, and nickel (Rouhi 1997). *Brassica juncea*, a mustard plant, will absorb lead, chromium, cadmium, nickel, and copper (Moffat 1995). Other plants will absorb copper, selenium, and cobalt, i.e., cobalt-60 (Rouhi 1997).

Some algae will absorb metal ions. For example, *Micrococcus luteus* can bind and remove strontium. The concentration of Sr-90 has been reported to be reduced from ppm to ppb very rapidly in some applications (Faison et al. 1990).

5.3.4 **Solidification (chemical/physical)**

Solidification technologies include in situ vitrification discussed above, grouting or cementing of soils and liquids either in place or for burial, and precipitation of inorganic ions from solution. Because a solidification technique may not change the character of the hazardous constituent, a demonstration that the waste form passes the RCRA TCLP test may be required before disposal in a RCRA-permitted facility can occur. Cement-based systems will bind metals and inorganic compounds, but do not immobilize organic compounds well. In fact, organic compounds inhibit the curing of concrete. Organic compounds adsorb on the faces of cement crystals and block the passage of water necessary for hydration. If the immobilized waste form later crumbles, organic compounds will be released. Grouting, of course, increases the volume of the waste form, as does above-ground vitrification. In situ vitrification does not increase the waste volume.

5.3.5 **Isolation (physical)**

Isolation technologies include placing grout curtains or sheet piling vertically in the soil around an existing waste site and disposing of waste in a new, RCRA-permitted hazardous waste disposal facility, i.e., landfill (see Sec. 6.1). In either case, a cap may be placed over the site to prevent precipitation from entering the waste form. The RCRA-permitted landfill will also have a liner/leachate collection system under the waste form. At least two types of cap are possible: a capillary barrier and an impervious barrier. The capillary barrier is a finely divided soil with pores that act as a capillary medium. Water from precipitation is held by the soil, like a sponge, until it can be evaporated or taken up by plant roots and transpired back to the atmosphere. The impervious barrier could be a geotextile, bentonite clay, or asphalt. Asphalt, however, degrades under ultraviolet light irradiation, so an asphalt cover must be placed under soil to keep out sunlight.

5.3.6 **Oxidation/Reduction (chemical)**

Oxidation processes (other than incineration) using hydrogen peroxide, hypochlorites, ClO_2 (made on site, dangerous stuff), electrolytic oxidation, ozone, etc., can be used separately or in combination with other techniques to oxidize organic compounds in both soil and water.

5.3.7 **Soil Washing (physical)**

Simply washing soil (solvent extraction) will remove soluble contaminants, which must then be removed from the water (or the solvent). If a solvent other than water is used, the excess solvent may remain trapped in the soil and have to be removed (by heating the soil, for example). Washing is one of the few ways of removing metal ions from soil.

5.3.8 **Soil Vapor Extraction (physical)**

Soil vapor extraction, sometimes called vacuum extraction, is an in situ technology similar to air stripping (discussed in Sec. 5.2.2), although the physical situation is different. A vacuum is applied through a well to soil in the vadose zone to remove volatile organic compounds (Figure 18). The technique will remove bulk amounts of organic compounds with high vapor pressures, but has not been shown to be effective in all kinds of soils down to small concentrations because of strong adherence of the organic compound to soil particles, particularly clays. A separate technology, such as incineration or charcoal adsorption, must be applied at the surface to remove the organic compound from the air stream.

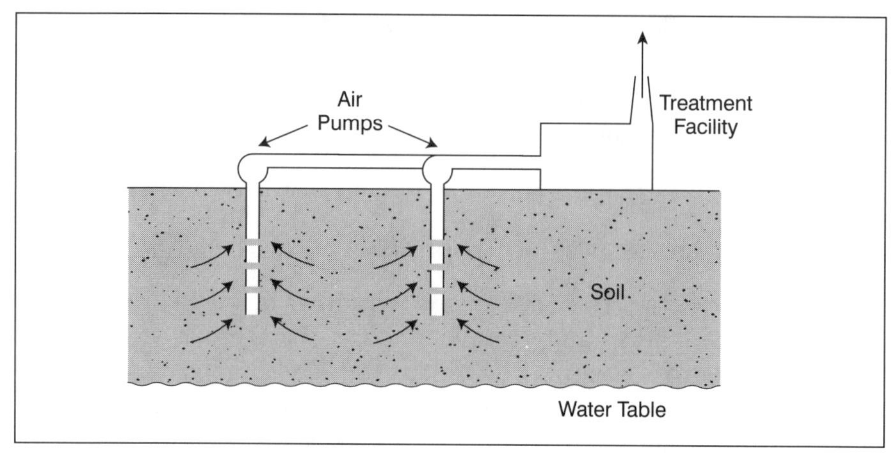

***Figure 18.** Soil vapor extraction.*

5.3.9 **Electroremediation (chemical)**

In situ electroremediation is carried out by inserting electrodes into the ground and applying a direct current voltage to the electrodes. In fortuitous cases, heavy metal cations such as arsenic, chromium, vanadium, and zinc will move through the soil to the negative electrode where they can be trapped and removed. Anions such as chromate will move to the positive electrode and may be captured there (Porter 1998). Care must be taken to avoid large currents that vitrify the soil.

5.3.10 **Combination Technologies**

Combinations of technologies are available, at least experimentally, for cleaning up liquids and slurries. For example, ultraviolet radiation, ozone, and hydrogen peroxide have been used to oxidize organic compounds. Catalysts such as TiO_2 are sometimes used to speed up reaction rates.

5.4 **DECOMMISSIONING NUCLEAR FACILITIES**

Decommissioning nuclear facilities requires special tools, equipment, and protection of workers because of the presence of radioactivity. At a minimum, workers must operate at a distance or behind shielding. Metal is cut under water with special saws and plasma arc torches because water is a good radiation shield. Some reactor piping must be cut remotely, or under water, because of high radiation dose rates. Dose rates and contaminated structures can be observed remotely using dosimeters and cameras mounted on robots. Robots can also carry out some decontamination activities. Highly radioactive wastes, including spent nuclear fuel, must be transported in casks, which are usually loaded and unloaded under water. Less radioactive solid wastes can be transported in shielded metal containers. Contamination control envelopes must be constructed for decommissioning activities that might involve releases of radionuclides to the atmosphere. These activities include spalling concrete and cutting metals in air.

5.5 MEDICAL WASTE CLEANUP TECHNOLOGIES

Medical waste is typically destroyed by incineration. Reusable equipment is cleaned by microwave radiation or steam sterilization.

CHAPTER 5 REFERENCES

Atlas, Ronald M. "Bioremediation," *Chemical and Engineering News*, p 32, April 3, 1995.

Environmental Protection Agency (EPA), *The Superfund Innovative Technology Evaluation Program, Progress and Accomplishments Fiscal Year 1989*, EPA/540/5-90/001, March 1990.

Environmental Science and Technology (ES&T). 31, 310A, July 1997.

Faison, Brendlyn et al. *Applied and Environmental Microbiology* 56 (12) 3649-56, 1990.

Hileman, Bette. "Times Beach Detox Nears End," *Chemical and Engineering News*, p 20, January 6, 1997.

Kopstein, Melvyn. "Science for Superfund Lawyers," 19 ELR 10388, September 1989.

Moffat, Anne Simon. "Plants Proving Their Worth in Toxic Metal Cleanup," *Science* 269, 302, July 21, 1995.

Porter, Amy. *Environment Reporter*, p 2370, March 6, 1998.

Rouhi, A. Maureen. *Chemical and Engineering News*, p 21, January 13, 1997.

Stone, Richard. "Oil-Cleanup Method Questioned," *Science* 257, 320, July 17, 1992.

CHAPTER 6

The Resource Conservation and Recovery Act

as amended by the Hazardous and Solid Waste Amendments (42 USC 690 et seq.)

RCRA was passed in 1976 as an amendment to the Solid Waste Disposal Act (SWDA) (enacted in 1965) and amended in 1984 by the HSWA. Solid wastes, garbage, hazardous wastes, and underground storage tanks are covered by RCRA, but only hazardous wastes and underground storage tanks are regulated directly. The major purposes of RCRA are to reduce or eliminate the generation of hazardous waste and to treat, store, and dispose of hazardous waste to minimize the present and future threat to human health and the environment (RCRA Sec. 1003). RCRA is a "cradle-to-grave" system (Figure 19) for managing hazardous waste that applies mainly to active facilities. RCRA requires a manifest system for generation, transportation, treatment, storage, and disposal of hazardous waste and requires permits for the treatment, storage, and/or disposal of hazardous waste (RCRA Subtitle C, Sections 3001-3019). HSWA institutes a land disposal ban program and a corrective action program. In addition, Subtitle I provides for the regulation of underground storage tanks containing "regulated substances." Regulated substances are petroleum and hazardous substances other than RCRA hazardous wastes. Significantly, RCRA also provides for "minimizing the generation of hazardous wastes...by encouraging...materials recovery . . . recycling and reuse." This gives rise to the important regulatory question of when a hazardous material becomes a regulated hazardous waste. Or,

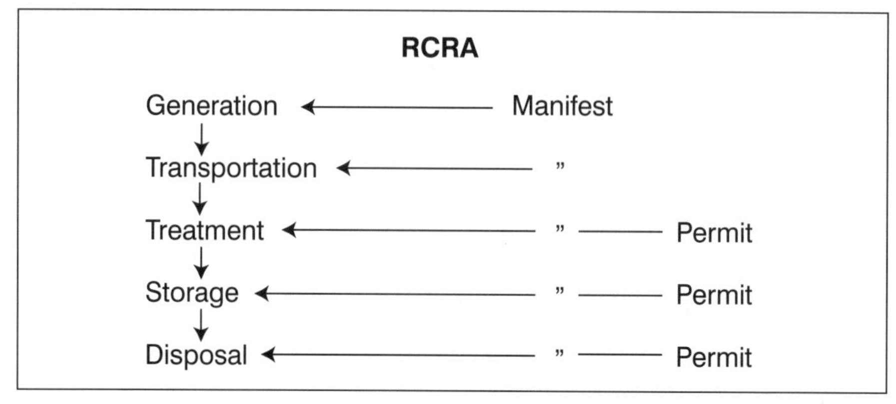

Figure 19. *Resource Conservation and Recovery Act.*

how long can a byproduct be stored before recycling without the by-product being considered a hazardous waste?

RCRA is administered either by the EPA or by the states. RCRA Sec. 3001 provides for the identification and either "listing" or "characterizing" of hazardous wastes. Characteristic wastes are corrosive, ignitable, reactive, or toxic wastes. RCRA Sec. 3002 requires generators of hazardous wastes to prepare manifests to track the generation, transport, treatment, storage, and disposal of the generated wastes. RCRA Sec. 3003 requires transporters of hazardous wastes to meet RCRA transportation requirements, as well as those of the Hazardous Materials Transportation Act (49 USC 1802 et seq.) and those in the Department of Transportation's Hazardous Waste Transportation regulations (49 CFR 171-179). RCRA Sec. 3004 requires the owners and operators of TSD facilities to comply with performance standards, groundwater monitoring requirements, and land-ban requirements prohibiting the disposal of untreated hazardous wastes in landfills. RCRA Sec. 3005 requires the owners and operators of TSD facilities to obtain RCRA permits and establishes "interim status." RCRA Sec. 3006 provides for state administration of the RCRA program. A waiver of sovereign immunity appears in RCRA Sec. 6001. RCRA Subtitle J, "Demonstration Medical Waste Tracking Program," provided for a brief demonstration program for tracking medical wastes.

The RCRA definitions of "solid waste" and "hazardous waste" are central to an understanding of RCRA and are to be distinguished from similar definitions in other laws, for example "hazardous substance," "pollutant," or "contaminant" in CERCLA and "pollutant" in the CWA.

> The term "solid waste" means any garbage, refuse, sludge, from a waste treatment plant or air pollution control facility and other discarded material ["discarded material" is not defined in RCRA], including solid, liquid, semisolid, or contained gaseous materials resulting from industrial, commercial, mining and agricultural activities and from community activities but does not include solid or dissolved material in domestic sewage, or solid or dissolved materials in irrigation return flows or industrial discharges which are point sources subject to permits under Sec. 112 of the Federal Water Pollution Control Act, as amended, or source, special nuclear, or byproduct material as defined by the Atomic Energy Act of 1954 as amended (RCRA Sec. 1004(27)).

> The term "hazardous waste" means a solid waste, or combination of solid wastes, which because of its quantity, concentration, or physical, chemical, or infectious characteristics may (A) cause, or significantly contribute to an increase in mortality or an increase in serious irreversible, or incapacitating reversible, illness, or (B) pose a substantial present or potential hazard to human health or the environment when improperly treated, stored, transported, or disposed of, or otherwise managed (RCRA Sec. 1004(5)).

In brief, "hazardous" in RCRA and in other environmental laws means injurious to your health or to the environment.

Wastes that are exempted from RCRA regulation, i.e., exempt or regulated elsewhere, include effluents regulated under the CWA; source, byproduct, and special nuclear material; irrigation return flows; household wastes; coal and other fossil fuel combustion wastes; drilling muds and brine used in oil, gas, and geothermal operations; mining wastes from extraction, treatment, and processing of ores and minerals; cement kiln dust; and wastes produced by small quantity generators (RCRA Sec. 3001(b)(3) and 40 CFR 261.4). At least one U.S. District Court (Idaho) has held, however, that mining wastes are hazardous substances under CERCLA (Idaho v. Hanna Mining Co. 1987).

The 1984 RCRA amendments (HSWA) shifted the emphasis on management of hazardous wastes from land disposal to treatment (see 40 CFR 268). Hazardous wastes must meet the treatment standard for the waste prior to land disposal or be disposed of in a facility from which there will be no migration of the hazardous constituents for as long as the waste remains hazardous. The treatment standards may be either levels or treatment methods (58 FR 48092, September 14, 1993).

The Federal Facility Compliance Act (FFCA) of 1992 broadens the waiver of sovereign immunity under RCRA and provides for state fines against the federal government for RCRA violations, although federal employees are not personally liable for civil penalties. The FFCA speaks specifically to mixed wastes and requires the federal government to prepare an inventory of mixed wastes, an inventory of treatment methods, and a plan for managing and disposing of mixed wastes. (Mixed wastes are mixtures of radioactive wastes and hazardous wastes. At the present time they are, for the most part, stored and not disposed of.)

6.1 EPA's RCRA REGULATIONS IN 40 CFR 260-281

EPA's Hazardous Waste Management regulations appear in 40 CFR 260-268 and 270-272. They apply to the generation, transport, treatment, storage, and disposal of hazardous wastes. In addition to the exemptions mentioned in Section 6.0, some exemptions apply to small quantity generators and to treatability studies. RCRA permits may be required by these regulations.

40 CFR 261
"Identification and Listing of Hazardous Waste."

The definition of hazardous waste in RCRA Section 1004(5) is not helpful with respect to specifically identifying a hazardous waste or the hazardous constituents of a mixture of wastes. The EPA's regulations in 40 CFR 261 are, however, more helpful. They specifically define and present lists of hazardous wastes. Hazardous wastes include both "listed" wastes (listed in 40 CFR 261) and "characteristic" wastes (defined in 40 CFR 261). Listed wastes are listed by chemical name or are listed as a waste stream from a specific chemical process. Examples include "hydrogen cyanide" (chemical name), and "plating bath residues from the bottom of plating baths from electroplating operations where cyanides are used in the process" (waste stream). Characteristic wastes are those that are ignitable, corrosive, reactive, or toxic. A toxic waste is one that fails the TCLP, which is described in 40 CFR 261. This test leaches the waste with acetic acid. If the leachate contains toxic compounds listed in 40 CFR 261.10 in sufficient concentration, then the waste itself is considered toxic. Recent regulations have brought the total number of toxic compounds listed in 40 CFR 261.10 to approximately 40 organic and inorganic compounds. The definitions of characteristic wastes are given in detail in 40 CFR 261.20-261.24. Stated briefly and somewhat generically, the definitions are:

- Corrosive hazardous wastes include aqueous wastes with a pH ≤ 2 or ≥ 12.5.
- Ignitable hazardous wastes include liquids with a flash point less than 60°C, wastes that undergo spontaneous combustion, and oxidizers.
- Reactive hazardous wastes include those that are normally unstable, react violently with water, generate a toxic gas when mixed with water, or are capable of detonation.
- Toxic hazardous wastes are those that fail the TCLP test.

40 CFR 262
"Standards Applicable to Generators of Hazardous Waste."

Regulations in 40 CFR 262 require EPA identification numbers for generators of hazardous waste and require manifests and detailed record keeping.

40 CFR 263
"Standards Applicable to Transporters of Hazardous Waste."

Regulations in 40 CFR 263 apply to transporters (both intra- and interstate) of hazardous wastes and require continued record keeping (upkeep of the manifest), as well as compliance with the DOT's hazardous waste transportation regulations in 49 CFR 171-179.

40 CFR 264
"Standards for Owners and Operators of Hazardous Waste Treatment, Storage, and Disposal Facilities."

Standards in 40 CFR 264 apply to owners and operators of TSD facilities that have RCRA Part B permits (permitted facilities). Facilities covered include containers, tank systems, surface impoundments, waste piles, land treatment facilities, landfills, incinerators, and miscellaneous units. Operation, closure, and post-closure periods are covered. Monitoring requirements, groundwater protection requirements, and corrective action requirements are specified. The regulations nominally contemplate a 30-year post-closure period. "Clean closure," however, means removal of all hazardous waste after which a post-closure period is not required. Otherwise, the 30-year post-closure period is required, which includes ground-water monitoring, maintenance, and monitoring of the waste containment systems. Corrective action means action that must be taken to clean up any portion of a hazardous waste facility from which hazardous wastes are migrating before a RCRA permit can be granted for continued operation of the facility. Corrective action requirements may be written into the RCRA permit.

Landfill requirements provide an example of the standards in 40 CFR 264. A new landfill (Figure 20) must have a liner/leachate collection system, which is a double-layer, impervious liner beneath the landfill with a piping system between the layers that can collect any leachate that penetrates the upper liner. Monitoring wells must be provided both upgradient and downgradient to detect any offsite release to groundwater. Closure requirements include some kind of cap on the waste form to inhibit or prevent moisture penetration. A cap could be an impervious barrier, or a capillary/evapotranspiration barrier, or both. A capillary/evapotranspiration barrier is one in which the barrier consists of fined-grained soils that hold moisture by capillary action until the moisture can be evaporated or transpired by plants to the atmosphere. Postclosure requirements include 30 years of continued well monitoring, which may be extended by the regulator.

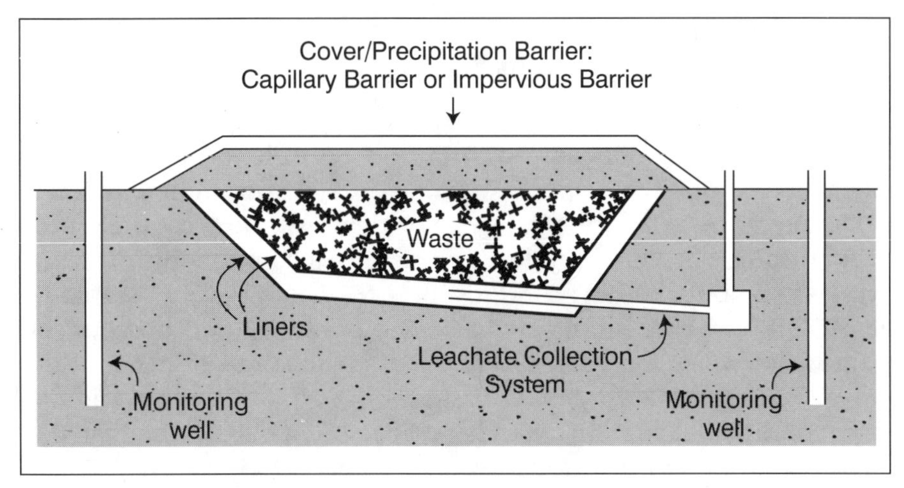

Figure 20. *Landfill showing RCRA requirements.*

Small quantity generators and laboratories engaged in small quantity treatability studies do not need a TSD permit. They must still meet RCRA requirements, however. The amount of hazardous wastes that can be handled at any one time in a treatability study laboratory were increased in 1994 (59 FR 8362, February 18, 1994).

40 CFR 265
"Interim Standards for Owners and Operators of Hazardous Waste Treatment, Storage, and Disposal Facilities."

Standards in 40 CFR 265 apply to the owners and operators of essentially the same hazardous waste facilities as those covered in 40 CFR 264, as well as thermal treatment units; chemical, physical, and biological treatment units; and underground injection wells, but not miscellaneous units, except that the facilities have interim status and do not have a RCRA permit. Interim status applies to existing facilities as they come under the purview of RCRA or new RCRA regulations. Interim status is attained by filing a short Part A RCRA permit application. The Part B permit application for status under 40 CFR 264 is a very lengthy application that fully describes TSD facility and its operation.

40 CFR 268
"Land Disposal Restrictions."

40 CFR 268 covers land disposal restrictions which, in general, prohibit land disposal of many hazardous wastes unless the wastes have been treated prior to disposal with the "best demonstrated available technology" (BDAT), or unless it can be demonstrated that the wastes will not migrate. This includes the hazardous components of mixed wastes. The bulk of EPA's land disposal restrictions were completed in 1990. These restrictions include a ban on the land disposal of solvent wastes (listed wastes, 1986), a ban on the land disposal of the so-called California list (listed wastes, 1987), a ban on the land disposal of dioxin-containing wastes (1988), and a ban on the land disposal of all other hazardous wastes (August 1988, June 1989, and August 1990). The California list includes cyanides, heavy metals, PCBs, halogenated organic compounds, and acids. (Note that PCBs and dioxins are also regulated under TSCA.) The regulations include treatment standards for hazardous wastes, i.e., allowable concentrations for land disposal, and acceptable treatment technologies.

Few facilities exist for treating or disposing of mixed wastes. And RCRA Sec. 3004(j) prohibits the storage of land-disposal prohibited wastes (including mixed wastes) except "for purposes of the accumulation of hazardous wastes as are necessary to facilitate proper recovery, treatment, or disposal."

EPA's acceptable RCRA LDR treatment technologies include incineration, biodegradation, charcoal adsorption, vitrification, encapsulation, grouting, and others. The technologies are different for different wastes. Dilution is, in general, not a solution, but can be used to deactivate corrosive, ignitable, or reactive, but not toxic, wastes.

40 CFR 264, 265, 270, and 271
Corrective Action for Solid Waste Management Units (SWMUs) at Hazardous Waste Management Facilities.
(mostly 40 CFR 264.100, 264.101, 264.552, and 264.553).

The corrective action regulations cover the cleanup of releases of hazardous wastes, usually to ground water, from existing hazardous waste management units as required by RCRA Sections 3004(u) and

3004(v). The corrective action protocol is very similar to (parallel in application to) the CERCLA protocol (Section 7.2 and Figure 21). The RCRA facility assessment (RFA) corresponds to the CERCLA preliminary assessment/site inspection (PA/SI), the RCRA facility investigation (RFI) corresponds to the CERCLA remedial investigation (RI), and the RCRA corrective measure study (CMS) corresponds to the CERCLA feasibility study (FS). The paperwork endpoint of the RCRA study process is a RCRA permit or permit amendment, and the paperwork endpoint of the CERCLA study process is a record of decision (ROD). The RCRA action process is the corrective measures implementation (CMI) which corresponds to the CERCLA remedial design/remedial action (RD/RA) process. Any RCRA TSD permit must address corrective action, as required. Conversely, any required corrective action will become a part of the RCRA TSD permit.

RCRA Corrective Action	**CERCLA Remedial Action**
RFA	PA/SI
	NPL
RFI	RI
CMS	FS
Permit	ROD
CMI	RD/RA

***Figure 21.** CERCLA/RCRA acronym comparison.*

40 CFR 270-272
RCRA Permit Program.

The regulations in 40 CFR 270-272 describe the RCRA permit program and the authorization of states to conduct the RCRA program.

40 CFR 280-281
Underground Storage Tanks.

EPA regulations in 40 CFR 280-281 (RCRA Subtitle I) apply to underground storage tanks containing regulated substances. A regulated substance includes petroleum and any hazardous substance regulated under CERCLA but not including any substance regulated as a hazardous waste under Subtitle C of RCRA, i.e., *hazardous substances* are included here except for *hazardous wastes.* EPA may authorize state regulation of underground storage tanks including a state permit. New underground storage tanks must have corrosion protection, which may include corrosion-resistant construction (fiberglass), cathodic protection, or a corrosion-resistant coating. Tanks containing hazardous wastes, i.e., tanks regulated under Subtitle C of RCRA, are regulated under 40 CFR 264 and 265.

40 CFR 258
"Criteria for Municipal Solid Waste Landfills."

The regulations in 40 CFR 258 "establish minimum national criteria . . . for all municipal solid waste landfill units." The criteria include siting, design, operating, monitoring, and closure requirements. Enforcement is carried out by the states.

6.2 OTHER INTERESTING RCRA ITEMS

Case law has held that Congress' power under the commerce clause of the U.S. Constitution allows Congress to impose outright prohibitions on the interstate movement of wastes. Congress may burden interstate commerce, in contrast to the states, which may not burden interstate commerce without Congressional approval. In other words, Congress can regulate interstate commerce, but the states cannot (Meitz 1990).

In Hazardous Waste Treatment Council v. State of South Carolina, the federal district court judge in early 1991 overturned South Carolina's ban on shipments of hazardous waste from neighboring states as a violation of the commerce clause. The court held that, although Congress can authorize the states to erect barriers to interstate commerce if expressly stated and unmistakably clear, neither RCRA nor

CERCLA provides states with the express authority to discriminate against out-of-state hazardous waste.

Similarly, Indiana was recently permanently enjoined from enforcing different procedures for out-state solid wastes than for in-state solid wastes.

Also, the Supreme Court in 1992 ruled that Alabama could not charge higher rates for the disposal in Alabama of hazardous wastes from outside the state than it could for the disposal of hazardous wastes from inside Alabama.

With respect to radioactive hazardous mixed wastes, treatment is limited and no disposal facilities are available. The mixed wastes are presently stored, although not legally, and cannot be legally disposed of in landfills without treatment.

As of August 9, 1993 (58 FR 42466), EPA has determined that fly ash, bottom ash, slag waste, and flue-gas emission-control waste from large-volume fossil fuel combustion facilities are not hazardous wastes under RCRA. This determination has its roots in the Bevill Amendment (RCRA Section 3001(b)(3)). Incinerator ash is, however, regulated under RCRA.

As an example of a RCRA permit, the Hanford Site-Wide RCRA permit was issued to DOE by EPA and the Washington Department of Ecology on August 29, 1994. Over 60 TSD units will eventually be included in the permit. This permit is an example of a single umbrella permit for a large site with many separate RCRA facilities. The permit for each separate site, as it is issued, will be added to the umbrella permit.

CHAPTER 6 REFERENCES

Meitz, Robert. "State Discrimination against Imported Solid Waste: Constitutional Roadblocks," 20 ELR 10383, September 1990.

CHAPTER 7

The Comprehensive Environmental Response, Compensation, and Liability Act

as amended by the Superfund Amendments and Reauthorization Act (42 USC 9601 et seq.)

CERCLA was passed in 1980 and amended by SARA in 1986. CERCLA provides for "liability, compensation, cleanup, and emergency response for hazardous substances released into the environment and the cleanup of inactive hazardous waste disposal sites" (preamble to CERCLA). More specifically, CERCLA provides 1) for remedial action at inactive or abandoned hazardous waste disposal sites (cleanup), 2) for removal action (also cleanup) of spills of hazardous substances, 3) for reporting releases to the environment of hazardous substances, and 4) for payment of damages for "injury to, destruction of, or loss of natural resources," i.e., natural resource damage assessments (NRDAs) (CERCLA Sec. 107(a)(4)(C)). A waiver of sovereign immunity appears in Sec. 120 of CERCLA.

The most visible part of CERCLA is the remedial action ("Superfund") part; and the usual distinction between CERCLA and RCRA is that CERCLA provides a remedial action program for past hazardous waste activities, while RCRA provides a regulatory program for present hazardous waste activities. Nevertheless, CERCLA includes three other very important provisions applicable to the release of hazardous substances, as distinguished from hazardous wastes. Hazardous substances are a wider universe of substances than are hazardous wastes. RCRA hazardous wastes are included in the universe of CERCLA hazardous substances.

The environment under CERCLA includes air, surface water, groundwater, land surface, and subsurface strata, but not biota (CERCLA Sec. 101(8)). CERCLA is directed at the release or substantial threat of release of any "hazardous substance" into the environment, or the release or threat of release into the environment of any "pollutant" or "contaminant" which may present an imminent and substantial danger to the public health or welfare (CERCLA Sec. 104). "Hazardous substance" is defined in general as any hazardous or toxic substance, pollutant, or chemical regulated under the CAA, CWA, TSCA, and/or RCRA (CERCLA Sec. 101 (14)), but not including petroleum. The complete list appears in 40 CFR 302. "Pollutant or contaminant" means any substance likely to cause death, disease, abnormalities, etc. (CERCLA Sec. 101 (33)). Note that RCRA is triggered by the existence of "hazardous wastes," while CERCLA is triggered by the release or substantial threat of release of any "hazardous substance." In other words, without the release or substantial threat of release of a hazardous substance, CERCLA cannot be invoked. Both "removal" actions and "remedial" actions are allowed under CERCLA. Removal actions refer to spills requiring immediate action, while remedial actions refer to longer-term cleanup of abandoned sites. No permits are required for CERCLA cleanup activities conducted entirely on site. Liability under CERCLA has been considered to be strict, joint and several, and retroactive, with narrow exceptions for the innocent land owner and financial institutions. Strict liability means that any person or organization that has had anything to do with the site or the hazardous waste can be held liable for the site or the waste. Joint and several liability means that any person or organization that has had anything to do with the site or the hazardous waste can be held completely liable for the site or the waste. Retroactive means that the law applies to actions that took place before the law was passed. Permanent remedial actions are favored under CERCLA (Sec. 107). The objective of CERCLA is protection of human health and the environment.

CERCLA is related to Section 311 of the Clean Water Act, which provides for the cleanup of oil and hazardous substances released to waters of the United States, under a national contingency plan (Light 1995).

7.1 EPA's CLEANUP REGULATIONS (40 CFR 300-302)

40 CFR 300
"National Oil and Hazardous Substances Pollution Contingency Plan."

EPA's CERCLA regulations in 40 CFR 300 apply to the cleanup of inactive hazardous waste disposal sites (remedial action), to the cleanup of hazardous substances released into the environment (removal action), to the cleanup of oil spills, and to the assessment of damages for injuries to natural resources. 40 CFR 300 is the so-called "National Contingency Plan" (NCP), which is originally from Sec. 311 of the CWA. The NCP provides for the development of a "National Priorities List" (NPL) of hazardous waste sites for which cleanup is mandated. Placement on the NPL triggers the remedial investigation/ feasibility study (RI/FS) process, which includes an RI/FS work plan, a remedial investigation of the site and the status of the site's releases or threatened releases, a risk assessment, an evaluation of no action, and a feasibility study (engineering evaluation) of cleanup methods (Figure 22). The last step of the study is a ROD, which records the cleanup method and the extent of cleanup (how clean is clean), which is dictated by the applicable or relevant and appropriate requirements (ARARs). The CERCLA ROD is not the same as a ROD issued following preparation of an environmental impact statement under the National Environmental Policy Act. Following publication of the CERCLA ROD, actual cleanup, i.e., remedial design/remedial action (RD/RA) begins (and may continue for years). Cleanup of a site without placement of the site on the NPL is not precluded under CERCLA and may proceed under CERCLA or under state law.

Contained within the NCP in 40 CFR 300.300-300.335 (Subpart D—Operational Response Phases for Oil Removal) are EPA's regulations covering removal of oil spills. These regulations are mandated more by the Clean Water Act and the Oil Pollution Act than by CERCLA.

Also contained within the NCP (40 CFR 300.400-300.440) are EPA's regulations for the removal of spills of hazardous substances.

Finally, EPA's regulations with respect to natural resource trustees and their duties are contained in 40 CFR 300.600-300.615. These regulations refer to 43 CFR 11, entitled "Natural Resource Damage

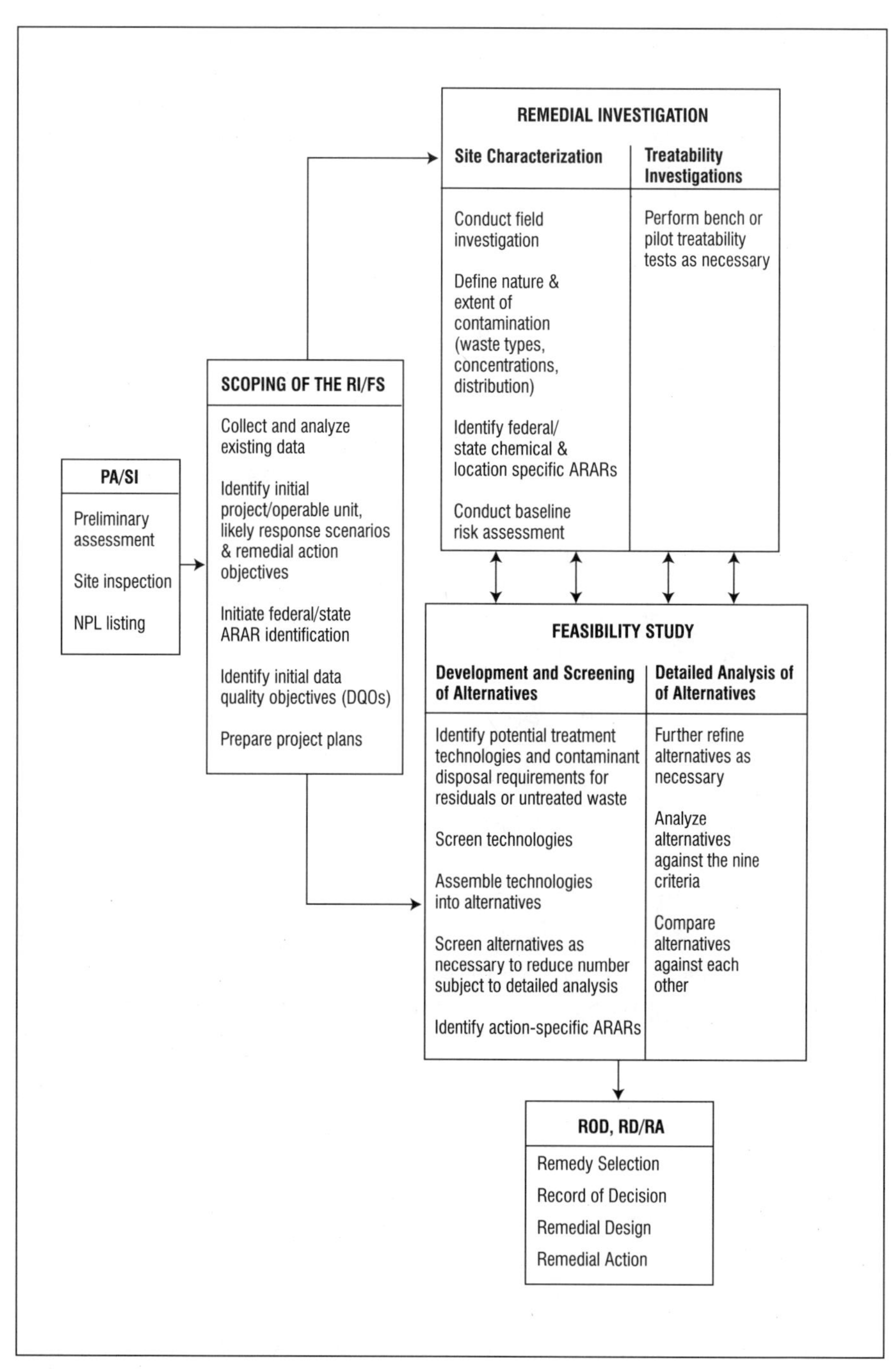

Figure 22. *RI/FS process.*

Assessments," which are the Department of the Interior's procedures for trustees to use to determine injuries to natural resources and damages to be assessed.

40 CFR 302
"Designation, Reportable Quantities, and Notification."

The EPA regulations in 40 CFR 302 list quantities of hazardous substances that must be reported if released to the environment. The list of hazardous substances in 40 CFR 302 is intended to include hazardous wastes designated under RCRA Sec. 3001 (40 CFR 261), hazardous air pollutants designated under CAA Sec. 112 (40 CFR 61), hazardous substances designated under CWA Sec. 311(b)(4) (40 CFR 116, 117), and toxic pollutants designated under CWA Sec. 307(a) (40 CFR 129).

40 CFR 355
"Emergency Planning and Notification."

These regulations contain a list of "extremely hazardous substances" under the EPCRA of 1986, which is a free-standing part of SARA. The list is a list of chemicals used in industrial processes that must be disclosed by industries to local units of government for emergency planning purposes.

7.2 THE RI/FS PROCESS (40 CFR 300.400-300.440)

The RI/FS process is outlined in Figure 22. It is important to note that the PA/SI may result in placement of the site or facility on the National Priorities List (NPL). Placement on the NPL then triggers the remedial investigation/feasibility study (RI/FS), leading to a ROD and to eventual cleanup of the facility or site through RD/RA. NPL activities are carried out under the supervision of EPA.

7.3 CERCLA RISK ASSESSMENT (40 CFR 300.430)

The CERCLA RI/FS process includes the preparation of a "site-specific baseline risk assessment to characterize the current and potential threats to human health and the environment that may be posed by contaminants migrating to ground water or surface water, releasing to air, leaching through soil, remaining in soil, and bioaccumulating in the food chain" (see Chapter 9). This baseline risk is the risk to the public of no action. The baseline risk assessment is then used to "help establish acceptable exposure levels for use in developing remedial alternatives," including no action. When ARARs (specific cleanup standards) are not available, exposure levels equivalent to a lifetime cancer risk of 10^{-6} may be used (as ARARs) for known or suspected carcinogens (40 CFR 300.430(e)). The extent to which an ARAR waiver (see Sec. 7.4) can be invoked when an ARAR exists, but the risk assessment shows no risk, remains to be seen.

It must be remembered that CERCLA is triggered by a release or a substantial threat of release of a hazardous substance. Thus, it would seem that if there is no release, no substantial threat of release, or no risk under the no action alternative, CERCLA cannot be invoked.

7.4 APPLICABLE OR RELEVANT AND APPROPRIATE REQUIREMENTS (40 CFR 300.400(g))

ARARs are "applicable or relevant and appropriate requirements" (40 CFR 300.5). They are cleanup standards defined in regulations promulgated by either federal or state agencies. Example ARARs include safe drinking water standards in 40 CFR 141, promulgated under the SDWA, or water quality standards promulgated by states under the CWA. They must be promulgated, they must be either "applicable" or "relevant and appropriate" (EPA says they cannot be both), and they must be met, unless one of six statutory waivers is invoked and approved by EPA. These waivers include a finding that "compliance with such requirement [ARAR] at that facility will result in greater risk to human health and the environment than alternative options," or that "compliance with such requirements is technically impracticable from an engineering perspective" (CERCLA Sec. 121(d)). ARARs include chemical-specific ARARs, location-specific ARARs, and action-specific ARARs. The most frequently mentioned chemical-specific ARARs are drinking water standards. Location-specific ARARs include historic preservation and species protection requirements. Action-specific requirements include such things as RCRA land-ban restrictions. For a discussion of some specific federal ARARs, see Chapter 10.

There are few federal cleanup standards for soils at the present time except for PCBs (40 CFR 761) and except for residual radioactivity at sites regulated by the U.S. Nuclear Regulatory Commission (10 CFR 20). Also, some states have soil cleanup standards that can be used as ARARs. Cleanup to background levels is the objective in some states.

The use of RCRA regulations as ARARs is at issue at the present time. For example, RCRA applies to rather concentrated quantities of wastes, while CERCLA wastes may be higher in volume and lower in waste content than RCRA wastes; thus RCRA treatment standards are not always particularly germane to CERCLA wastes. RCRA permit requirements and RCRA land disposal restrictions may make it difficult to dispose of some CERCLA wastes.

7.5 THE NATIONAL ENVIRONMENTAL POLICY ACT AND CERCLA

NEPA requires federal agencies to prepare an environmental impact statement (EIS) for any major federal action significantly affecting the quality of the human environment. Many CERCLA cleanup actions fit this definition. However, the extensive environmental documentation requirements and the broad public participation requirements mandated by CERCLA for the cleanup of sites on the NPL are duplicative of the NEPA requirements for federal sites. This duplication has led to extensive discussions among federal agencies, including the Council on Environmental Quality (CEQ), which administers NEPA, and the EPA, with respect to integrating CERCLA and NEPA requirements. The question is whether federal agencies should be required to prepare an EIS for a CERCLA action when the requirements of NEPA and CERCLA are so similar. For the time being, it would seem that the preparation of an environmental impact statement (EIS) under NEPA is required for the CERCLA cleanup of a federal site on the NPL, although it must be pointed out that some CERCLA actions on federal sites are now proceeding without an EIS. Not long ago, the CEQ and DOE said both were required, but the Department of Justice said that the RI/FS process is equivalent to an EIS and therefore both are not required. CEQ or Congressional action may be required to eliminate the duplication.

7.6 NATURAL RESOURCE DAMAGE ASSESSMENTS (40 CFR 300 AND 43 CFR 11)

CERCLA provides (in Sec. 107(a)(4)(C)) for the payment of monetary damages for "injury to, destruction of, or loss of natural resources, including the reasonable costs of assessing such injury, destruction, or loss resulting from such a release [of a hazardous substance] . . ." "Natural resources" means "land, fish, wildlife, biota, air, water, ground water, drinking water supplies, and other such resources belonging to, managed by, held in trust by, appertaining to, or otherwise controlled by the United States . . . , any state or local government, any foreign government, any Indian tribe . . ." (CERCLA Sec. 101(16)). Damages are to be paid to the U.S. Government, any state, or any Indian tribe for injury, destruction, or loss of natural resources. At the same time, agencies of the federal government, states, and Indian tribes are also liable for damages, except that federally permitted actions, actions specifically identified in an environmental impact statement as irreversible or irretrievable commitments of natural resources where a permit or license authorizes such commitment, and actions occurring before the passage of CERCLA in 1980 are not subject to the assessment of damages. Courts have held that the preferred standard for measuring natural resource damages is the cost of completely restoring or replacing a damaged resource. For monetary damages to be assessed under the NRDA clause, an injury must have occurred, the injury must be shown to be caused by a release of a hazardous substance, and the injury must be quantifiable.

Oil spill NRDAs are covered under the Oil Pollution Act of 1990 and the oil spill regulations in 40 CFR 300.

The Department of the Interior (DOI) has promulgated regulations in 43 CFR 11—Natural Resource Damage Assessments—that supplement "the procedures established under the NCP, 40 CFR part 300, for the identification, investigation, study, and response to a discharge of oil or release of a hazardous substance, [and provide] a procedure by which a natural resource trustee can determine compensation for injuries to natural resources . . ."

7.7 OTHER INTERESTING CERCLA ITEMS

In 1992, 32,000 sites were being evaluated for possible CERCLA remediation. A total of 75,000 sites had been estimated as possibly requiring remedial action at a cost of $1 trillion for the most stringent cleanup. In addition, there were possibly 37,000 sites where RCRA corrective action might be required (Abelson 1992). By 1995, EPA's inventory consisted of 15,600 high-priority sites with over 1,300 specifically on the NPL. At that time only 346 sites were considered to be cleaned up (CEQ 1996). The major CERCLA problems are, of course, soil and ground water contaminated with radioactivity, heavy metals, and organic solvents.

The Community Environmental Response Facilitation Act (1992), which amends CERCLA Sec. 120(h), allows the federal government to sell uncontaminated portions of a federal site before the entire site is cleaned up, i.e., before all hazardous wastes and petroleum products are removed from the entire site.

Four Superfund sites in western Montana encompass 50,000 acres along 140 miles of the Clark Fork River and its tributaries. Contaminants include As, Cd, Pb, Cu, and Zn. Most Superfund sites are somewhat smaller in size!

CHAPTER 7 REFERENCES

Abelson, Philip H. "Remediation of Hazardous Wastes," *Science*, 255, 901, February 21, 1992.

Council on Environmental Quality (CEQ). *Environmental Quality 25th Anniversary Report*, CEQ, 1996.

Light, Alfred R. "Deja Vu All Over Again?: A Memoir of Superfund Past," *Natural Resources and Environment* 10, 29, Fall 1995.

CHAPTER 8

Cleanup of Oil Spills: The Oil Pollution Act

Oil is excluded from the CERCLA definition of hazardous substance and from the CWA definition of hazardous substance. However, before the passage of CERCLA, the CWA in Sec. 311 provided for the cleanup of discharges of both oil and hazardous substances (as defined under the CWA) from a vessel, offshore facility, or onshore facility into or on the navigable waters or adjoining shorelines, or that may affect natural resources, or that may be a substantial threat to the public health or welfare. Thus Sec. 311 of the CWA provides for cleaning up both offshore and onshore discharges or spills of oil. Reporting these spills of hazardous substances and oil is also required by Sec. 311 of the CWA (see Smith 1993).

The OPA (33 USC 2701 et seq.) is closely related to Sec. 311 of the CWA and was enacted as a result of the Exxon Valdez oil spill. The OPA is also closely related to CERCLA. The OPA includes broadened provisions for the cleanup of oil spills on navigable waters and shorelines. Sections 1002(b) and 1006 of the OPA provide for natural resource damage assessments for oil spills, just as Sec. 107(f) of CERCLA provides for natural resource damage assessments for spills of CERCLA hazardous substances. The OPA also includes provisions for improved safety of oil transport, for example, double hulls for tank vessels (Sec. 4115).

EPA's regulations pertaining to the cleanup of oil spills can be found within the CERCLA regulations in 40 CFR 300, "National Oil and Hazardous Substances Pollution Contingency Plan," specifically in 40 CFR 300.300-300.350 (Subpart D, "Operational Response Phases for Oil Removal").

Petroleum in underground storage tanks is also regulated under RCRA Subtitle I (RCRA Sec. 9001-9010). The EPA regulations in 40 CFR 280-282 provide for detecting, reporting, and cleaning up releases of petroleum from underground storage tanks and also provide for standards of performance for new underground storage tanks. Household tanks are exempted from Subtitle I.

EPA's regulations pertaining to the management of used oil can be found in 40 CFR 279, "Guidelines for the Management of Used Oil." These regulations apply to the handling of used and recycled oil. They apply specifically to generators, transporters, and processors of oil, to burning oil for energy, and to the use of oil as a dust suppressant.

CHAPTER 8 REFERENCES

Smith, Al J., Jr. "Inland Oil Spills in the Nineties: Avoiding the Confusion," *Environment Reporter*, 2509, January 22, 1993.

CHAPTER 9

Risk Assessment

Risk is considered in this book because it is necessary to understand what EPA means by regulating exposure to some hazard so that the risk of cancer is in the range of 10^{-4} to 10^{-7} and because the CERCLA regulations require that a baseline risk assessment be carried out during the RI/FS process. EPA's definition of acceptable risk is an additional lifetime (70-year) ***individual*** cancer risk to the potential maximum exposed individual of 1 in 100,000 or 10^{-5}. Individual risks, however, cannot be determined except in very few situations such as the individual risk of death, which is 1. What is determined, at least in theory, is the risk to a number of individuals in a large population. That is, the risk to a given individual cannot be identified, but the risk to a number of individuals in a large population can, in theory, be determined.

At the outset, risk as used above is a probability. Risk is a number between zero and one. A risk of 10^{-4} means a probability of 0.0001, which is a number between zero and one. A risk of 10^{-4} also means that one person in 10,000 persons will be affected, not that a given person will be affected. It is important to recognize that risk really applies to populations of persons and not to individuals, which is how risks are determined in the first place, i.e., numbers of persons affected in a large population.

To determine a risk, the number of consequences of some event must be known, and the total population under consideration must be known. The risk is the number of consequences divided by the population under consideration. Consequences can be the number of deaths, injuries, illnesses, etc. caused by the event, and an event can be normal annual operations or an accident. When the number of consequences and the population under consideration are known, the calculation of risk is a simple matter. When the consequences must be calculated based on a mathematical or experimental model, the determination of risk is less straightforward. Much time and effort goes into determining the number of consequences where good statistical data are not available.

To complicate matters, risk is sometimes reported as the number of consequences from an event and the frequency of the event, but the population under consideration is not mentioned at all. This, then, is not a risk in the sense that risk is a probability or number between zero and one. This is, however, a perfectly acceptable way of reporting the situation. Frequency can be a number of events per year (accidents, for example), or a steady event (a release, for example) throughout the year.

9.1 THE POLITICS OF RISK

Risk can mean "actual risk," "calculated risk," or "perceived risk." An "actual risk" is one that is measured, i.e., a risk for which good statistical data are available, for example the risk of automobile accidents. "Calculated risk" is one in which the number of consequences must be determined partly by experiment or observation, but mostly by mathematics. "Perceived risk" is one that is in the eye of the beholder.

Confusion exists in the literature between the term "risk assessment" and the term "risk analysis" (Cohrssen and Covello 1989). This is not important except to note that different authors use the terms somewhat differently. For our purposes, we will consider risk assessment to include the narrower term risk analysis.

Risk assessment may be thought of as consisting of three parts: risk analysis, risk communication, and risk management. Risk analysis is the science component in which one attempts to determine a risk. Risk communication is the social science or public policy component, which includes risk perception and risk acceptability. And risk management combines both science and public policy. The public policy part of risk management is, of course, an EPA or state regulation. The science part is the basis for setting the regulation and the basis for developing the protocol to meet the requirements of the regulation.

9.2 RISK

Risk is the probability of loss, injury, sickness, or death. Risk is a pure number between zero and 1. The individual risk of dying is 1. The risk of dying this year in the U.S. (in a population of a large number of people) is 2 million divided by 250 million, or 0.01 (10^{-2}). (The population of the United States is approximately 250 million and approximately 2 million people die each year.) Each year, 500,000 people die from cancer, therefore the probability (population risk) of dying from cancer this year is 500,000 divided by 250 million, or 0.002 (2×10^{-3}). Fifty thousand people die each year from automobile accidents, therefore the probability (population risk) of dying in an automobile accident this year is 50,000 divided by 250 million, which is 0.0002 (2×10^{-4}).

Over a lifetime, where the probability of dying is 1, the probability of dying from cancer is 500,000 divided by 2 million, or 0.25 (2.5×10^{-1}). Over a lifetime the probability of dying from an automobile accident is 50,000 divided by 2 million, or 0.025 (2.5×10^{-2}).

Up to this point, things are relatively simple. The number of consequences and the population under consideration are fairly accurately known because they are based on reported statistics. Further, the risks are simply calculated by dividing the consequences (numbers of deaths) by the total number of persons in the population being considered. These results can, in turn, be used to predict numbers of consequences in similar populations. But these risks do not involve the intermediate step of determining the number of consequences from some event, such as a release of a hazardous substance, that results in the consequences. This complicates things.

The determination of the numbers of consequences from an event, such as a release of radioactivity, is not always easy. These determinations involve a pathway analysis, as well as a dose-response relationship, which is a human response (death or illness) caused by a given dose. Complete human epidemiological data are not available with which to determine numbers of consequences. And without human epidemiological data, extrapolations of questionable validity must be made from animal data at high rates to animal responses at low rates, and from animal responses at low dose rates to human responses at low dose rates. Stated a little differently, we know the risk of dying from all forms of cancer and even the risk of dying from different forms of cancer, but it is much more difficult (almost impossible) to know the risk of dying from cancer caused by, say, exposure to a specific chemical.

An intermediate case where substantial human data are available is exposure to radiation. Here, the BEIR (Biological Effects of Ionizing Radiation) folks (BEIR 1990) have been doing their thing since 1970 (and before), which is to study disease and death from ionizing radiation based on information from exposures and medical records, including those from the Japanese atom bomb survivors. The BEIR studies are based on human epidemiology, i.e., on human populations that are as large as possible, so that there is less need to rely on animal data. The 1990 BEIR average population risk prediction (dose-response relationship) for men and women is approximately 800 cancer deaths per million person-rem for acute doses and approximately 600 cancer deaths per million person-rem for more chronic doses.

Based on these statistics, it is interesting to compute the risk to the U.S. population of dying from background radiation. Six hundred cancer deaths per person-rem × 0.3 rem per year background dose × 250 million persons equals 45,000 cancer deaths per year. The risk (probability) of dying later from background radiation (delivered this year) is 45,000 divided by 250 million or 0.0002 (2×10^{-4}). The risk of dying over a lifetime from background radiation is 45,000 divided by 2 million or 0.02 (2×10^{-2}). These formulas apply to populations. Attempts to apply them to individual risks break down. For example, an acute dose of 400 rem to an individual, which is almost always fatal, computes to an individual probability of death of 0.3 for that one person, i.e., 800 cancer deaths per million person-rem × 400 rem equals 0.3 deaths for that one person, instead of 1.0 death. (For an acute dose this large, however, death actually results rapidly from causes other than cancer.)

This example, where the statistics are reasonably good, suggests the desirability of relying on good human epidemiological data, rather than on animal data, and reveals the unfortunate result when individual risk is computed based on population risk. It gets worse when attempts are made to determine human risks based on animal data in those cases where human data are not available, which is almost all other cases. There are two reasons for this. One, the human response may not be the same as the response in laboratory animals, i.e., potentially invalid extrapolations must be made from animals to humans; for example, in 1987, only 20 of the 600 known animal carcinogens were known also to be human carcinogens (Ames 1987). Two, extrapolations must be made from high animal doses (often the "maximum tolerated dose"—800 mg/kg-day in the example below) to low animal doses based on mathematics rather than on observed responses (the extrapolation may be from a high-dose data point directly back to zero) (Abelson 1990).

Consider the following experiment, which could be performed on rats to evaluate a dose-response relationship:

Lifetime Daily Dose (mg/kg-day)	Lifetime Incidence of Liver Cancer in Rats	Risk (Lifetime Probability of Liver Cancer)	Potency Factor (Risk/Dose in kg-day/mg)
0	0/50	0.0	—
150	0/50	0.0	0
300	10/50	0.2	6.7×10^{-4}
500	30/50	0.6	12.0×10^{-4}
800	40/50	0.8	10.0×10^{-4}

In this experiment, groups of 50 rats are given daily lifetime doses of an imaginary chemical as stated in the first column. The incidence of liver cancer at death is recorded in the second column. The risk and a potency factor are calculated in columns three and four. Risk versus dose is plotted in Figure 23.

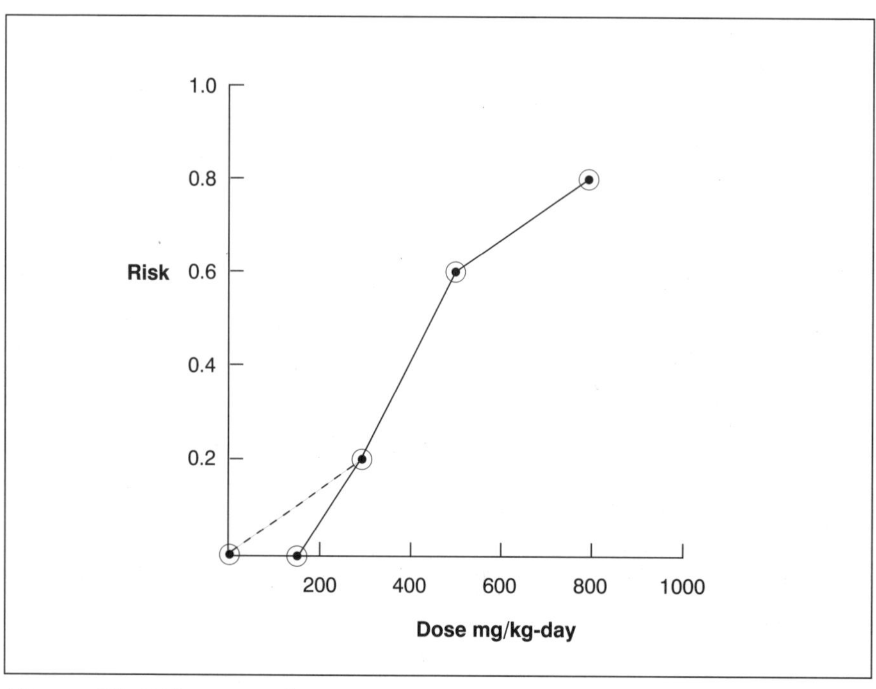

Figure 23. *Risk versus dose.*

The data are then applied to a human population. For example, assume that if each member of a human population receives a dose of 1 mg/kg-day, what is the population risk? If one believes the animal data as plotted in Figure 23, then the answer is clearly zero (i.e., based on the data in the graph, there is a threshold). However, conventional wisdom says that there is no threshold dose for cancer, therefore to compute a population risk at low dose, resort must be made to mathematical hocus-pocus. One choice is to use the potency factor (which one should you choose?). This is a straight (linear) extrapolation from a point on the graph back to zero and not following the curve back to the abscissa. The smallest of these (6.7×10^{-4}—dashed line on Figure 23) computes to a population risk of 6.7×10^{-4} kg-day/mg × 1 mg/kg-day = 6.7×10^{-4}. Other mathematical extrapolations (the Weibull distribution, for example) lead to risks from 6.0×10^{-5} to 1.9×10^{-10} for the human ingestion of 1 mg/kg-day. If, for some reason, it is important to limit the human population risk to, say, 1×10^{-6}, then the model chosen becomes very important because it, in turn, limits the daily human intake.

All of this is made more complicated by many other factors. For example:

- There are various pathways for harmful substances to reach humans
- The amount received internally must be correlated with the ambient concentration
- There may be an intermediate species such as fish with a bio-accumulation factor of its own
- The matter of no threshold for carcinogens and a threshold for noncarcinogens is still under debate
- In some cases, severity of response is more important than incidence
- Effects on different organs are important.

Potency factors (slope factors), as calculated by the EPA, are available from EPA's Integrated Risk Information System (IRIS).

Recent data (Ames and Gold 1990) indicate that many chemicals that cause cancer in animals do not damage genes, i.e., are not mutagenic. It is thus possible that the large doses simply kill animal cells (perhaps not unlike large doses of radiation), thereby stimulating the replacement of these cells by rapidly growing and dividing new cells. Rapidly dividing cells are more susceptible to mutation than are quiescent cells. Thus the cancer-causing mutation may occur in the rapidly dividing new cells rather than in the cells damaged by the high chemical dose. The cancer may arise from cell proliferation rather than from the genotoxic effects of the chemical. Quoting directly from Ames and Gold: "you cannot understand mutagenesis (and therefore carcinogenesis) without taking mitogenesis into account and that at high doses chronic mitogenesis can be the dominant factor" (Ames and Gold 1990).

The 1990 amendments to the Clean Air Act include the following directive to EPA: The Administrator of EPA shall enter into appropriate arrangements with the National Academy of Sciences to conduct a review of . . . "risk assessment methodology used by the Environmental Protection Agency to determine the carcinogenic risk associated with exposure to hazardous air pollutants . . ." This review could result in substantial changes to the way EPA determines risk.

9.3 RISK-BASED STANDARDS

Environmental standards set by EPA are either based on risk (risk-based standards) or on available technology (technology-based standards). A technology-based standard can be met (presumably) by existing technology. A risk-based standard may or may not have the technology available to meet the standard. SDWA standards are supposedly risk-based standards, while prevention of significant deterioration (PSD) and CWA discharge standards are technology-based standards.

CHAPTER 9 REFERENCES

Abelson, Philip H. "Incorporation of New Science into Risk Assessment," *Science* 250, 1497, December 14, 1990.

Ames, B.N. *Regulatory Toxicology and Pharmacology*, 7, 379, 1987.

Ames, Bruce N., and Gold, Lois S. "Too Many Rodent Carcinogens: Mitogenesis Increases Mutagenesis," *Science* 249, 970, August 31, 1990, and *Science* 250, 1498, December 14, 1990.

Biological Effects of Ionizing Radiation (BEIR). *Health Effects of Exposure to Low Levels of Ionizing Radiation*, BEIR V, National Research Council, National Academy Press, Washington, D.C., 1990.

Cohrssen, John J., and Covello, Vincent T. *Risk Analysis: A Guide to Principles and Methods for Analyzing Health and Environmental Risks*, Council on Environmental Quality, Washington, D.C., 1989.

CHAPTER **10**

Applicable or Relevant and Appropriate Requirements

Environmental laws and regulations apply on their own, independently of CERCLA; but they may also serve as ARARs under CERCLA, i.e., they may serve as CERCLA cleanup standards. In this section, a brief summary of federal environmental laws and regulations that may serve as ARARs is presented. State environmental laws and regulations may also serve as ARARs, but are not discussed here, except to note that most state environmental laws and regulations complement, duplicate, or are derived from federal environmental laws. In some cases applicable state regulations are more stringent than federal regulations.

Just as "all things are connected" environmentally, all things are connected legally, i.e., CERCLA is not an isolated environmental law. The environmental laws and regulations discussed here may have a direct bearing on CERCLA as ARARs.

Environmental standards and permit requirements usually appear in regulations and not in the laws themselves. Thus, in this section, both laws and regulations are discussed and cited.

10.1 FEDERAL ENFORCEMENT, STATE ENFORCEMENT, AND SOVEREIGN IMMUNITY

The activities of the federal government are ordinarily not subject to regulation by the states (U.S. Constitution, Article V Section 2: Federal Supremacy). Congress, however, has created specific exceptions in most environmental laws. These waivers of sovereign immunity appear in the Clean Air Act, Clean Water Act, Safe Drinking Water Act, Resource Conservation and Recovery Act, and the Comprehensive Environmental Response, Compensation, and Liability Act. The waivers of sovereign immunity provide either for the delegation of environmental regulatory authority over federal facilities to the states, usually by the EPA, or provide for the outright regulation of federal facilities by the states without delegation by the EPA. Courts have usually held that the waiver must be clear and unmistakable to be valid.

CERCLA, therefore, applies to federal facilities either directly (enforcement by EPA) or through its waiver of sovereign immunity (enforcement by states). However, once a federal site or facility is placed on the CERCLA National Priorities List, the EPA (the federal government) and not the state manages the CERCLA cleanup process. Nevertheless, both federal and state laws and regulations may be used as ARARs in the CERCLA cleanup of a federal facility.

10.2 CLEAN AIR ACT (42 USC 7401 et seq.)

The major purpose of the Clean Air Act (CAA) is "to protect and enhance the quality of the Nation's air resources" (CAA Sec. 101(b)). The CAA in its present form was enacted in 1970 and was amended in 1977 and in 1990. It deals both with stationary sources of air pollutants (factories, refineries, power plants, etc.) and with mobile sources of air pollutants (airplanes and automobiles); provides for the establishment of "National Ambient Air Quality Standards" (NAAQS, Sec. 109), which include standards for sulfur dioxide, particulate matter, carbon monoxide, ozone, oxides of nitrogen, and lead (these six are called "criteria pollutants"); provides for the creation of "State Implementation Plans" (SIP, Sec. 110), to be approved by EPA to allow states to impose controls on existing stationary sources to reduce emissions to meet NAAQSs; provides for "New

Source Performance Standards" (NSPS, Sec. 111), to be applied to new or modified industrial stationary sources; provides for the "Prevention of Significant Deterioration" (PSD, Sec. 160-190) of air quality from new or modified sources in regions that already meet the NAAQSs; and provides for the creation of "National Emission Standards for Hazardous Air Pollutants" (NESHAPs, Sec. 112), which include radon-222 from underground uranium mines and from uranium mill tailings, beryllium, mercury, vinyl chloride, radionuclides from DOE facilities, benzene, asbestos, arsenic, and more up to possibly as many as 189. Three major permits or authorizations are established in the CAA: PSD, NESHAP, and nonattainment area. These three permits are evolving into a single comprehensive air quality permit under the Clean Air Act Amendments of 1990, which will include toxic sources (NESHAPs), attainment area sources (PSD), nonattainment area sources, and sources subject to NSPSs. This single permit will include some facilities not subject to air quality permits before 1990. Also, a waiver of sovereign immunity is included in the CAA (Sec. 118). Chlorofluorocarbons (protection of the ozone layer) and acid rain are also covered in the CAA.

The NAAQSs are not directly enforceable. Rather, they set the standards on which other enforceable requirements, such as emission limitations and permit requirements, are based. These standards can be used as ARARs should facilities such as incinerators be needed for cleanup. Permits were not required for existing sources before 1990, unlike existing sources under the CWA. Now, CAA permits will apply to all major sources of air pollutants as well as to sources of hazardous air pollutants.

The 1990 amendments to the CAA do not change the basic framework of the CAA, however they do include a list of 189 hazardous air pollutants which EPA must consider for regulation. These pollutants and their management could become the subject of ARARs in the future for any cleanup process that involves releases to the atmosphere. CAA requirements also apply to RCRA facilities such as incinerators. For example, emission requirements for incinerators can be found in 40 CFR 264.

EPA's Air Quality Regulations appear in 40 CFR 50-99. Again, the regulations, rather than the CAA itself, need to be consulted for potential ARARs.

10.3 **CLEAN WATER ACT (33 USC 1251 et seq.)**

The purpose of the Clean Water Act (CWA) is to "restore and maintain the chemical, physical, and biological integrity of the Nation's waters" (CWA Sec. 101(a)). The CWA was enacted in 1972 as the Federal Water Pollution Control Act (FWPCA) and was amended in 1977 as the CWA. The "waters" referred to in the CWA purpose are largely surface and not ground waters. Some protection of ground water is afforded by the Safe Drinking Water Act. The CWA provides for water quality standards (Sec. 303), for effluent standards for a long list of industries (Sec. 301), for the listing and regulation of toxic pollutants (a short list of toxic organic pollutants subject to NPDES permits including aldrin, dioxin (TCDD), DDT, and PCBs) (Sec. 307), for the listing and regulation of oil and hazardous substances (closely related to CERCLA) (Sec. 311), for the regulation of publicly owned treatment works (POTWs), for state certification of activities requiring a federal permit (Sec. 401), for National Pollutant Discharge Elimination System (NPDES) permits (Sec. 402), and for dredged or fill material discharge permits (Sec. 404). The 1977 amendments established three categories of pollutants: toxic pollutants, including an initial list of 129 chemicals (Sec. 307(a)), conventional pollutants (Sec. 304(a)(4)), and nonconventional pollutants (Sec. 301(g)(1)). Conventional pollutants under the CWA include BOD, suspended solids (SS), fecal coliform bacteria, and acidity as measured by pH. (BOD is the amount of oxygen necessary to decompose organic material in water.) Nonconventional pollutants include ammonia, chlorine, and iron. Sec. 316 covers thermal discharges specifically and permits state regulation of thermal discharges. A waiver of sovereign immunity appears in Sec. 313.

Unlike the old CAA, where requirements are established to meet the NAAQSs, requirements under the CWA are directed at uniform discharge standards from both existing and new sources even if the discharges do not result in violation of water quality standards. It is interesting that pollutants under the CAA and Safe Drinking Water Act are chemicals, but CWA pollutants include more than just chemicals (see definition of "pollutant" below). Any limitation of a pollutant discharge to waters of the United States has the potential to become an ARAR.

EPA's Water Quality Regulations appear in 40 CFR 100-140 and 400-501.

Definitions of interest under the CWA are given below. These definitions can be confusing. For example, a hazardous substance as defined under CERCLA includes a hazardous substance as defined under the CWA. What is important here is that a limitation on the discharge of any CWA hazardous substance or pollutant into waters of the United States during a CERCLA cleanup action may constitute an ARAR under CERCLA. The CERCLA manager needs to be aware of that possibility.

> A "hazardous substance" is defined as "such elements and compounds [other than oil] which, when discharged in any quantity into or upon the navigable waters of the United States or adjoining shorelines...present an imminent and substantial danger to the public health or welfare, including, but not limited to, fish, shellfish, wildlife, shorelines, and beaches" (CWA Sec. 311(b)(2)(A)). These substances may be included in an NPDES permit.
>
> The term "pollutant" means dredged soil, solid waste, incinerator residue, filter backwash, sewage, garbage, sewage sludge, munitions, chemical wastes, biological materials, radioactive materials, heat, wrecked or discarded equipment, rock, sand, cellar dirt, and industrial, municipal, and agricultural waste discharged into water" (CWA Sec, 502(6)).

However, the U.S. Supreme Court on June 1, 1976, held that "pollutants" subject to regulation under the CWA do not include source, byproduct, or special nuclear materials in Train v. Colorado PIRG.

> The term "toxic pollutant" means those pollutants, or combinations of pollutants, including disease causing agents, which after discharge and upon exposure, ingestion, inhalation or assimilation into any organism, either directly from the environment or indirectly by ingestion through food chains, will . . . cause death, disease, behavior abnormalities, cancer, genetic mutations, physiological malfunctions (including malfunctions in reproduction) or physical deformation, in such organisms and their offspring" (CWA Sec. 502(13)).

10.4 SAFE DRINKING WATER ACT (42 USC 300f et seq.)

The Safe Drinking Water Act (SDWA) was enacted in 1974 and amended in 1986 and provides for the regulation of public drinking water systems from both surface and ground water sources (Sec. 1411), for the establishment of national standards for levels of contaminants in drinking water (Sec. 1412), for the regulation of underground injection wells (five classes) (Sec. 1421), for the identification of sole source aquifers (approximately 50 now) (Sec. 1427), and for the state protection of wellhead areas (Sec. 1428). Sections 1421, 1427, and 1428 provide some protection of ground water. Under Sec. 1413, a state may be delegated primary enforcement responsibility by EPA for public drinking water systems. A waiver of sovereign immunity appears in Sec. 1447 of the SDWA. This waiver is with respect to federally owned or maintained public drinking water systems and with respect to federal underground injection activities that may endanger drinking water.

EPA has the authority to designate a groundwater aquifer as a "sole source aquifer" if the EPA determines that the aquifer is the sole or principal source of drinking water for an area and that its contamination would create a significant hazard to human health. When this happens, federal financial assistance is barred to activities that could contaminate the aquifer and create a significant hazard to human health. The designated aquifer could cover a very large area.

EPA's SDWA regulations appear in 40 CFR 141-149. The EPA's drinking water standards, which are called maximum contaminant levels (MCLs) and maximum contaminant level goals (MCLGs), are specific ARARs under CERCLA Sec. 121. Regulations for use of underground injection wells could also be CERCLA ARARs if underground injection of water treated in a CERCLA cleanup action is contemplated.

10.5 RESOURCE CONSERVATION AND RECOVERY ACT (42 USC 6901 et seq.)

The Resource Conservation and Recovery Act (RCRA) and its regulations may serve as action-specific ARARs under CERCLA. For example, RCRA land disposal restrictions may apply as ARARs to CERCLA hazardous wastes disposed of on site. This onsite disposal is complicated by the fact that the RCRA regulations usually contemplate concentrated waste forms, while CERCLA often deals with more dispersed waste forms. This sometimes makes it inconvenient to apply RCRA regulations as ARARs. EPA's RCRA regulations appear in 40 CFR 260-281. See Chapter 6 for a complete discussion of RCRA.

10.6 TOXIC SUBSTANCES CONTROL ACT (15 USC 2601 et seq.)

The Toxic Substances Control Act (TSCA) provides for the testing and regulation of chemical substances entering the environment. EPA may regulate the manufacture, use, distribution, and disposal of both new and old chemical substances that enter the environment. Manufacturers must notify the EPA before producing a new chemical substance, the EPA can then require testing before the substance enters the environment. If a chemical substance is an unusual risk, the EPA may restrict its manufacture and use. TSCA supplements the CAA and the CWA and applies to chemicals that are not necessarily considered wastes. Sections 201-215 cover asbestos in schools and in public and commercial buildings. Sections 301-311 cover indoor radon abatement.

EPA's regulations in 40 CFR 700-799 implement TSCA and, in particular, regulate asbestos (also regulated under the CAA in 40 CFR 61), polychlorinated biphenyls (PCBs), and dioxins, the latter two being regulated under the CAA, CWA, and RCRA, as well. Concentration standards for PCBs in soils in the TSCA regulations can be ARARs under CERCLA.

10.7 SPECIES PROTECTION ACTS

Significant species protection acts under federal law include the Endangered Species Act (ESA) (16 USC 1531 et seq.), the Fish and Wildlife Coordination Act (16 USC 661 et seq.), the Bald and Golden Eagle Protection Act (16 USC 668 et seq.), and the Migratory Bird Treaty Act (16 USC 703 et seq.). The ESA provides for the designation and protection of threatened or endangered species and provides for the designation of critical habitat for certain threatened or endangered species. The other acts also provide protection for wildlife and for certain birds.

Species protection regulations appear in 50 CFR 10-24, 219-229, 402, and 450-453. These regulations are administered by the U.S. Fish and Wildlife Service and the National Marine Fisheries Service. The species protection laws and regulations may be very important location-specific or action-specific ARARs under CERCLA, depending on the species that might be present at or near a CERCLA cleanup site. The CERCLA cleanup manager is advised to consult the U.S. Fish and Wildlife Service, the National Marine Fisheries Service, and any state counterparts about possible impacts on protected species.

10.8 HISTORIC, SCENIC, AND CULTURAL PRESERVATION ACTS

Federal historic, scenic, and cultural preservation acts include the National Historic Preservation Act (NHPA) (16 USC 470 et seq.), the Archaeological Resources Protection Act (16 USC 470aa et seq.), the Archaeological and Historic Preservation Act (16 USC 469-469c et seq.), the American Antiquities Act (16 USC 431 et seq.), the American Indian Religious Freedom Act (42 USC 1996), the Native American Graves Protection and Repatriation Act (25 USC 3001 et seq.), and the Wild and Scenic Rivers Act (16 USC 1274 et seq.).

Historic, scenic, and cultural preservation regulations appear in 36 CFR 800, 36 CFR 18, 36 CFR 60, 36 CFR 63, 36 CFR 79, 36 CFR 296, 25 CFR 261, 43 CFR 3, and 43 CFR 7. Regulations of the National Historic Preservation Act appear in 36 CFR 800, 36 CFR 18, 36 CFR 60, 36 CFR 63, and 36 CFR 79; regulations of the American Antiquities Act appear in 25 CFR 261 and 43 CFR 3; and regulations of the Archaeological Resources Protection Act and the American Indian Religious Freedom Act appear in 36 CFR 296 and 43 CFR 7. These regulations apply to the protection of historic and cultural properties, including both existing properties and those discovered during excavation and construction. The state historic preservation officer (SHPO) has been delegated substantial authority under federal law for the local administration of historic preservation. Historic preservation acts and regulations may have significance as action- and location-specific ARARs under CERCLA. The CERCLA cleanup manager is advised to consult the state historic preservation officer about historic or cultural resources that may be impacted by the cleanup action.

10.9 INDIAN TREATY RIGHTS

Treaties ratified by Congress are the supreme law of the land along with the Constitution and federal law (U.S. Constitution Art. 6). Many treaties were written by the U.S. Government with Indian tribes in the 1800s and early 1900s. These treaties give specific rights to tribes, and these rights must be protected and may not be violated during any CERCLA cleanup activity.

10.10 NUCLEAR MATERIALS AND NUCLEAR WASTE HANDLING ACTS

The Atomic Energy Act (42 USC 2011 et seq.), the Low-Level Radioactive Waste Policy Act (42 USC 2021b et seq.), and the Nuclear Waste Policy Act (42 USC 10101 et seq.), while not environmental laws per se, contain provisions under which environmental regulations applicable to radioactive materials may be or have been promulgated. These acts apply both to government-owned radioactive materials managed and regulated by the U.S. Department of Energy (DOE) and regulated by EPA in some cases, and to commercially owned radioactive materials regulated by the Nuclear Regulatory Commission (NRC) and/or by agreement states. An agreement state is one which has an agreement with the NRC to handle some of the NRC's regulatory functions.

The NRC regulations appear in 10 CFR 0-199. These regulations may have significance as ARARs under CERCLA whenever radionuclides are involved (for example, NRC's residual radioactivity regulations in 10 CFR 20).

10.11 OTHER SIGNIFICANT ENVIRONMENTAL REGULATIONS

The following regulations are not covered earlier in this section, but may apply to CERCLA cleanups as ARARs.

33 CFR 322
"Permits for Structures or Work in or Affecting Navigable Waters of the United States."

Structures and work in navigable waters require a U.S. Army Corps of Engineers (COE) permit under Section 10 of the Rivers and Harbors Act (33 USC 401 et seq.). Navigable waters are defined by the COE in 33 CFR 329.

33 CFR 323
"Permits for Discharges of Dredged or Fill Material into Waters of the United States."

The discharge of dredged or fill material into waters of the United States requires a COE permit under Section 404 of the CWA. The EPA has the authority to veto a Section 404 permit on environmental protection grounds (40 CFR 230-233). States can veto Sec. 404 permits on the grounds that the activity will cause a violation of the state's water quality standards. Waters of the United States are defined by the COE in 33 CFR 328.

49 CFR 171-179
"Hazardous Materials Regulations."

Department of Transportation (DOT) regulations in 49 CFR 171-179, promulgated under the Hazardous Materials Transportation Act (49 USC 1802 et seq.), apply to the handling, packaging, labeling, and shipment of hazardous materials, including radioactive wastes. These regulations apply to the transport of RCRA hazardous wastes and to the offsite transportation of any hazardous wastes from CERCLA cleanup activities.

10.12 ENVIRONMENTAL PERMITS

Major federal environmental permits are listed below. These permits may be administered either by a federal agency or by a state agency, depending upon whether or not the state has received federal approval to issue the permit. Furthermore, all of these permits apply to federal agencies and, again depending on federal approval, may be administered by the states. Finally, some of these permits may be required as adjuncts to hazardous waste handling under RCRA or hazardous waste cleanup under CERCLA. It is to be noted once again that activities conducted entirely on site during a CERCLA cleanup action do not require a permit. Offsite actions related to a CERCLA cleanup may, however, require one or more of the following environmental permits.

CAA: PSD permit (40 CFR 52)

CAA: NESHAP authorization (40 CFR 61 and 63)

CAA:	New or modified major source in a nonattainment area permit (40 CFR 52)
CAA:	After 1990, a single air quality permit including all of the above (40 CFR 70 and 71)
CWA:	NPDES permit (40 CFR 122)
CWA, RHA:	COE permits (33 CFR 322-323 and 40 CFR 230-233)
SDWA:	Underground injection control (UIC) permit (40 CFR 144)
RCRA:	RCRA TSD permit (40 CFR 264-265)
RCRA:	UST permit (40 CFR 280-281)

10.13 STATE ARARs

Promulgated state regulations may serve as ARARs.

10.14 NUMERICAL STANDARDS FOR PROTECTION OF THE PUBLIC

Numerical standards for protection of the public from releases to the environment have been set by the EPA, by other federal agencies including the NRC, and by the states. The standards appear in air quality, water quality, solid waste, and radioactivity regulations. These standards may have significance as ARARs under CERCLA, including standards for residual contamination, i.e., how clean is clean? Some examples are presented here.

40 CFR 61 and 63 "National Emission Standards for Hazardous Air Pollutants."

The standards in 40 CFR 61 apply to the release of hazardous substances to the atmosphere, including benzene, asbestos, vinyl chloride, arsenic, and some radionuclides. Standards set for the 189 hazardous air pollutants listed in the CAAA of 1990 appear in 40 CFR 63. An example is the standard in 40 CFR 61.92 for emissions of radionuclides to the atmosphere from DOE facilities:

> Emissions of radionuclides [other than radon 220 and 222] to the ambient air from Department of Energy facilities

shall not exceed those amounts that would cause any member of the public to receive in any year an effective dose equivalent of 10 mrem/yr.

40 CFR 141
"National Primary Drinking Water Regulations."

MCLs and MCLGs in 40 CFR 141 apply directly to CERCLA water cleanups as ARARs and indirectly to releases of radionuclides and hazardous substances into water, to the extent that the releases impact community water systems. For example, EPA's MCL with respect to radionuclides is:

> The average annual concentration of beta particle and photon radioactivity from man-made radionuclides in drinking water shall not produce an annual dose equivalent to the body or any internal organ greater than 4 millirem/year.

Also, MCLs in community water systems of 20,000 picocuries per liter of tritium and 8 picocuries per liter of Sr-90 are specified in 40 CFR 141. 40 CFR 141 also specifies maximum concentrations of an increasing number of chemical contaminants in drinking water. For example, the drinking water standard (MCL) for nitrate as nitrogen in water is 10 mg/L.

10 CFR 0-199
U.S. Nuclear Regulatory Commission Regulations.

The NRC regulations in 10 CFR 0-199 apply to radioactive materials and nuclear facilities owned by commercial establishments and may apply as ARARs to any CERCLA cleanup involving radionuclides. Regulations include 10 CFR 20, "Standards for Protection against Radiation," and 10 CFR 61, "Licensing Requirements for Land Disposal of Radioactive Waste." NRC's residual radioactivity regulation may be found in 10 CFR 20.1402:

> "A site will be considered acceptable for unrestricted use if the residual radioactivity that is distinguished from background radiation results in a total effective dose equivalent to an average member of the critical group that does not exceed 25 mrem (0.25 mSv) per year, including that from groundwater sources or drinking water . . . "

CHAPTER **11**

Liability for Violations of Federal Environmental Laws

Most federal environmental laws now contain provisions for administrative, civil, and criminal penalties for violations of the law. Furthermore, most of these federal environmental laws also contain federal waivers of sovereign immunity that make these sanctions and penalties available against responsible agencies, officers, and employees of the federal government, as well as against responsible federal government contractors, their officers, and their employees.

Administrative penalties (fines and orders) can be levied by administrative agencies, but these may be appealed to the courts. A civil penalty is a fine for injuries, which a court can levy against a company or against persons responsible for the injuries. Courts can issue orders to stop illegal activities. Civil actions include both citizens' suits, where the penalty (a fine) goes into the U.S. Treasury, and toxic tort suits, where damages are sought by individuals allegedly harmed by an action. A civil penalty can include both compensatory damages and punitive damages. Criminal penalties can be levied by a court against companies, against individuals responsible for the violation, and against corporate officers. Criminal penalties can include both a fine for the company (federal agency) and for individuals, and a jail term for individuals.

11.1 CIVIL ENFORCEMENT

Civil enforcement (with fines and penalties) is available under the CAA, CWA, CERCLA, EPCRA, OSHA, RCRA, SDWA, and TSCA.

11.2 CITIZENS' SUITS

Citizens' suits can be brought under the CAA, CWA, CERCLA, EPCRA, RCRA, SDWA, and TSCA.

11.3 CRIMINAL SANCTIONS

Criminal charges can be filed for violations of the CAA; CWA; CERCLA; EPCRA; FIFRA; Marine Protection, Research, and Sanctuaries Act; OSHA; RCRA; SDWA; and TSCA.

As an oddity, federal employees are criminally liable under the CWA, but are exempt from civil liability within the scope of their employment.

CHAPTER **12**

Siting Hazardous Waste Facilities

Siting large energy facilities has been a major issue in the United States since the early 1970s. This issue has both a technical component and a sociopolitical component. The technical component includes engineering, environmental, and economic issues that usually have solutions; while the sociopolitical component more often than not is the NIMBY issue, which has no solution. ("NIMBY" stands for "not in my back yard.") Facilities that are unsatisfactory for my back yard are equally unsatisfactory for anyone else's back yard.

It is now virtually impossible to site a large energy facility in this country. Hazardous waste facilities and radioactive waste facilities are not far behind. Almost no progress has been made in siting commercial low-level radioactive waste burial grounds since the passage of the Low-Level Radioactive Waste Policy Act in 1986. And not much more progress has been made in siting hazardous waste facilities since the passage of RCRA. Construction and operation of the Ward Valley commercial low-level waste burial ground in California has been tied up in controversy and litigation for years (Sec. 3.4). Siting and operation of the commercial hazardous waste incinerator in East Liverpool, OH, were also tied up for years.

A review of the Washington State dangerous waste facility siting regulations, which contain dangerous waste siting criteria, is instructive.

The Washington State Department of Ecology (WDOE) "Standards for Siting Hazardous Waste Treatment and Disposal Facilities" contain criteria for siting dangerous waste management, treatment, storage, and disposal facilities, including container storage, tank storage, incinerators, landfills, waste piles, and surface impoundments. The technical requirements are directed at protecting the environment; and they are, for the most part, not difficult to meet by straightforward engineering practices. But the technical requirements also lead directly to sociopolitical issues that will not have any solutions. Consider, for example, the criteria for siting landfills, but keep in mind that two of Washington's climatic regions are the wet, heavily populated, and highly developed region around Puget Sound and the very dry (desert), lightly populated, and relatively undeveloped region in southeast Washington.

WDOE standards for siting hazardous waste landfills include the following:

- the landfill must be one quarter mile from surface water, dwellings, and wells
- the landfill must be one-quarter mile from prime farmland, wetlands, parks, and wilderness areas
- the landfill must be 50 feet above ground water
- the landfill must not be within the 500-year flood plain.

Other criteria include provisions for seismic risk and hydrogeologic studies before siting and ground-water monitoring and protection after siting. There are, however, no criteria that address sociopolitical considerations such as siting landfills close to the facilities that produce the hazardous wastes.

A good case could be made that the criteria bias the location of suitable landfill sites away from the wet and populated Puget Sound area toward the dry and less populated southeast area of the state. It would seem that, through accident of climate, population, and development, the State of Washington criteria will mitigate the NIMBY effect for the Puget Sound area and exacerbate it for the southeast area.

CHAPTER 13

Conclusion

Two of the "laws" or principles considered early in the text are relevant in all facets of hazardous waste management and are worthy of emphasis at the end of our study. These are "all things are connected" and "don't let it get out in the first place."

"All things are connected" is certainly true of hazardous waste management (and it is true of the web of environmental law that exists in this country). Almost any pollutant released to the air, water, or even deposited on or into the ground can eventually make its way to the other two media; and more often than not with deleterious effects, if in a large enough concentration. The only action that can prevent this is "don't let it get out in the first place." Any serious practitioner of hazardous waste management will have no trouble keeping these principles in mind because they are inescapable.

Index

All federal agencies are listed under their complete names (for example, you would look for U.S. Environmental Protection Agency under U).

Figures are denoted by an *f* following the page the figure appears on. If you look at an entry and expect to see a figure and don't, the page the figure is on is part of a page range and thus does not appear separately.

CERCLA=Comprehensive Environmental Response, Compensation, and Liability Act
RCRA=Resource Conservation and Recovery Act

CERCLA=Comprehensive Environmental Response, Compensation, and Liability Act
RCRA=Resource Conservation and Recovery Act

CERCLA=Comprehensive Environmental Response, Compensation, and Liability Act
RCRA=Resource Conservation and Recovery Act

CERCLA=Comprehensive Environmental Response, Compensation, and Liability Act
RCRA=Resource Conservation and Recovery Act

CERCLA=Comprehensive Environmental Response, Compensation, and Liability Act
RCRA=Resource Conservation and Recovery Act

CERCLA=Comprehensive Environmental Response, Compensation, and Liability Act
RCRA=Resource Conservation and Recovery Act

CERCLA=Comprehensive Environmental Response, Compensation, and Liability Act
RCRA=Resource Conservation and Recovery Act

CERCLA=Comprehensive Environmental Response, Compensation, and Liability Act
RCRA=Resource Conservation and Recovery Act

CERCLA=Comprehensive Environmental Response, Compensation, and Liability Act
RCRA=Resource Conservation and Recovery Act